Trent's Last Case

Edmund Clerihew Bentley was born in 1875 in Shepherd's Bush, London, the son of a civil servant. At St Paul's School, he befriended G. K. Chesterton and in 1890 invented the 'clerihew'—a humorous verse-form in which nonsensical biographies are composed in four unscanned lines. He studied history at Oxford, where he was President of the Union and captain of Merton Boat Club. After graduating in 1898, he studied law and drifted into journalism, first as a columnist and leader-writer for the *Daily News* (1901–12), and then with the *Daily Telegraph* (1912–34), also making occasional contributions to *Punch* and other magazines. His first book of clerihews, *Biography for Beginners*, illustrated by Chesterton, appeared in 1905, and was followed by *More Biography* (1929) and *Baseless Biography* (1939). In 1981 a notebook was discovered in St Paul's School library containing many hitherto unpublished clerihews by Bentley, Chesterton, and their schoolfriends, dating from some time before 1893; this was published as *The First Clerihews* (1982).

In 1913 Bentley published his first and most successful detective story, *Trent's Last Case* (at first under the title *The Woman in Black* in America). He revived the character of the detective Philip Trent in *Trent's Own Case* (with H. Warner Allen, 1936) and in *Trent Intervenes* (short stories, 1938), and later wrote an adventure novel, *Elephant's Work* (1950). Bentley published his memoirs in 1940 as *Those Days*. He returned from retirement to the *Daily Telegraph* as a literary critic during the 1940s. In his later years he was depressed by the wartime destruction of his house, and turned to drink. He died in 1956. He had married Violet Boileau in 1901; of their three children, one, Nicolas Bentley, became a celebrated illustrator.

Trent's Last Case

E. C. Bentley

Introduced by

Chris Baldick

Oxford   New York

OXFORD UNIVERSITY PRESS

1995

Oxford University Press, Walton Street, Oxford OX2 6DP

Oxford New York
Athens Auckland Bangkok Bombay
Calcutta Cape Town Dar es Salaam Delhi
Florence Hong Kong Istanbul Karachi
Kuala Lumpur Madras Madrid Melbourne
Mexico City Nairobi Paris Singapore
Taipei Tokyo Toronto
and associated companies in
Berlin Ibadan

Oxford is a trade mark of Oxford University Press

British Library Cataloguing in Publication Data
Data available

Library of Congress Cataloging in Publication Data
Data available
ISBN 0–19–282422–8

1 3 5 7 9 10 8 6 4 2

Typeset by Best-set Typesetter Ltd., Hong Kong
Printed in Great Britain
by Biddles Ltd Guildford & Kings Lynn

OXFORD POPULAR FICTION

General Editor   Professor David Trotter

Associate Editor   Professor John Sutherland
Department of English, University College London

Amongst the many works of fiction that have become bestsellers and have then sunk into oblivion a significant number live on in popular consciousness, achieving almost folkloric status. Such books possess, as George Orwell observed, 'native grace' and have often articulated the collective aspirations and anxieties of their time more directly than so-called serious literature.

The aim of the Oxford Popular Fiction series is to introduce, or re-introduce, some of the most influential literary myth-makers of the last 150 years—bestselling works of British and American fiction that have helped define a new style or genre and that continue to resonate in popular memory. From crime and historical fiction to romance, adventure, and social comedy, the series will build up into a library of books that lie at the heart of British and American popular culture.

CONTENTS

INTRODUCTION

Trent's Last Case is one of the classics of English detective fiction, but it became a classic in spite of itself. In some accounts, it has been credited as the first work of a new 'Golden Age' of country-house murder-mysteries, heralding the later arrivals of Margery Allingham, Nicholas Blake, and Dorothy L. Sayers. Its author, the humorist and newspaper columnist Edmund Clerihew Bentley, on the other hand, never meant it so seriously at all, and was pleasantly bemused by the status his novel soon acquired. From his schooldays, when he became the precocious inventor of a ludicrous form of unscanned ditty known in his honour as the clerihew, Bentley cultivated a cheerful impertinence as the basis of his literary endeavours, avoiding any suspicion of solemnity or even serious ambition. It may be taken as some sort of omen that one of his first juvenile clerihews took the reigning master of the detective story as its victim:

> Conan Doyle
> Ought to be boiled in oil.
> He makes no reference to gnomes
> In 'The Adventures of Sherlock Holmes'.

The possibility of a more light-hearted rival to the hermit of Baker Street may even have occurred to the irreverent schoolboy at this time in 1892 or 1893. To this seminal early poem, at any event, can be traced the evolution of Bentley's challenge to the old regime in detective fiction. He would write the gnomes back in.

Bentley's conscious purpose in writing *Trent's Last Case* was to make a new kind of detective story, unburdened by what he felt were the absurdities and the already tiresome conventions of the form. He believed that his book was, as he put it, 'not so much a detective story as an exposure of detective stories'—a kind of travesty which undermined the cult of the infallible master-sleuth. As he recalled later in his memoirs, *Those Days* (1940), the basis of his plan for the proposed novel was a dissatisfaction with two aspects of Sherlock Holmes and of the lesser detectives based upon him. The first cause for complaint was Holmes's exaggerated unreality or eccentricity as a character; the second was the high seriousness and austerity associated with his apparently superhuman mental powers. Bentley set himself the challenge in these terms: 'It should be possible, I thought, to write a detective story in which the detective was recognizable as a human being, and was not quite so much the "heavy" sleuth.' This objective occurred to him in 1910, when he was working as a journalist on the staff of the *Daily News* off Fleet Street, and he pursued it over the next few months, as an amusing test of his invention—'as a lark', he later said. In his daily walks from his house in Hampstead to the *Daily News* offices, he ran the possibilities through his mind, working and reworking the twists of an ingenious plot. At an early stage he jotted down some of the essential fictional ingredients he would require: 'a millionaire—murdered, of course; a police detective who fails where the gifted amateur succeeds; an apparently perfect alibi; some fussing about in a motor car . . . ' and 'a crew of regulation suspects, to include the victim's widow, his secretary, his wife's maid, his butler, and a person who had quarrelled openly with him.' Clearly, in attempting a new kind of detective story, Trent was not abashed to begin with several well-established clichés of the genre, even including the unimaginative police inspector, obviously based upon Conan Doyle's Inspector Lestrade. At some early stage he decided too, with reluctance, that a love-interest would be expected in a modern novel, and thus would need to be woven in. Rather than extend the cast of characters especially for this purpose, Bentley decided to set himself the additional challenge of making plausible the 'extreme absurdity' of a romantic attraction between the detective and the murdered man's widow.

After several weeks of ambulatory calculation, Bentley went on to the writing of the novel, spending his spare evenings for at least six months standing at a desk at home (he preferred not to write sitting down), sometimes working into the small hours of the mornings. The effort involved is reflected in the book's title: Bentley decided that he could not sustain the labour of another detective novel, and so this, his own first 'case', would also be his last—although, more than twenty years later, he abandoned this resolution. The completed draft, called at this point *Gasket's Last Case*, was sent to the London publishing house of Duckworth in 1911, in the hope of winning a fifty-pound prize offered at that time for the best first novel submitted in that season. While awaiting the verdict, Bentley fell into a chance conversation with an energetic young American called Douglas Z. Doty, of the New York publishers Century, who expressed interest in the story, hinting that it could be worth far more than fifty pounds. Tipped off by a Duckworth insider that his book was not going to win their prize, the creator of Gasket hastily reclaimed his script and took it to Doty, who was about to sail back to New York. Doty read it on the boat, and eventually cabled over an offer of $500 (just over £100 at 1912 rates) for the American rights. His company insisted on changing the hero's unappetizing surname, which they felt must become a monosyllable; thus the detective became 'Trent'. At the same time, Century wanted a change of title, and they brought out the first American edition, in March 1913, as *The Woman in Black*, which Bentley regarded as a silly interference. Meanwhile, Bentley's second approach to a British publisher, this time the firm of Thomas Nelson, had borne fruit: the popular Scottish novelist John Buchan was an active partner in the firm, and his recommendation secured *Trent's Last Case* a place in Nelson's series of two-shilling novels. This British edition also appeared in March 1913, earning Bentley a further £75 as an advance on what later proved to be more significant royalties. The author took special delight in the speed with which foreign-language rights were sold and translations commissioned, at first in Danish, Swedish, and Italian, then in French, German, Polish, Russian, and eventually Gaelic.

We have seen that Bentley was aiming, in this novel, to create a new kind of detective story, while still employing a good number of stock ingredients. The principal novelty in his plan was to be found in the portrayal of the detective, who was to be given a lighter, more 'human' character than those of Poe's Auguste Dupin or Doyle's Sherlock Holmes. By contrast with the now famous Holmesian eccentricities (introspective pipe-smoking, violin-playing, the cocaine habit, elaborate disguises, obscure monographs, lack of sexual interest), Philip Trent is given a set of positively 'normal' traits, including a genial good humour which prevents him from taking himself too seriously, a fondness for country ale and plain English food, and even a certain capacity to conduct heterosexual courtship. However, just because he is more likely to be found with a glass of beer in his hand than with a syringe of cocaine, Philip Trent should not be identified immediately with the average healthy 32-year-old English gentleman of his day. Trent retains one important feature of the nineteenth-century detective, and that is his 'artistic' temperament. He is a successful painter who is cajoled unwillingly into a return to amateur detective work, and clearly would rather be back in his studio, completing the painting which this case has interrupted. He is credited with a love of poetry, and indeed frequently resorts to poetical quotation, not without cheerfully mangling the verses of his favourite Romantic authors. The conflict between the cultured taste represented by Trent and the materialistic or 'philistine' mentality of the late Mr Manderson is central to the book's scale of implied values. For now, we can merely note that Philip Trent is neither a lavish eccentric, of the kind represented by Holmes or by Christie's Hercule Poirot, nor a merely 'ordinary' sample of red-blooded English manhood, but a compromise between these positions. 'Imagination is my forte,' he tells Calvin Bunner, emphasizing his artistic nature, but then he immediately pays tribute to the more useful experience and methodical diligence of his humbler rival, Inspector Murch. Trent wears his imagination and his culture casually, lightening it with laughter, and never slipping into a Holmesian tone of superiority. He lacks the maniacal *intensity* of a Holmes or a Dupin, as Bentley stresses both his 'loosely built' physique and his 'large and loose' cultural range.

The most important new characteristic given to Philip Trent is an endearing human fallibility. Bentley's strongest objection to the 'heavy' solemnity of the older detectives was prompted by their apparently superhuman infallibility, and the cult of absolute mastery that they evoked, notably in the Sherlock Holmes cycle. It is true that even Sherlock Holmes can be found making minor mistakes in the handling of his cases, for which he upbraids himself extravagantly, all the better to underline his perfectionism; but, although he may let the occasional criminal slip away (upon his own admission he has been beaten four times), his intellectual grasp of the case will, in the end, always be definitive. Bentley's radical innovation was the creation of a detective whose brilliant deductions could be flawed by oversights and misplaced assumptions. It would not be fair to the reader to 'give away' in this introduction the precise nature or consequences of the hero's fallibility in this novel, but nobody's enjoyment of the tale will be spoiled by being told simply that Trent is susceptible to error, that his clear vision has a small 'blind spot' which endangers his success both as a detective and as a lover. This has, on the whole, just the effect that Bentley had hoped it would have: a stronger sense of the detective's full humanity, supported further here by the many evidences of his humour, his vivacity, and his appetite for food, drink, and conversation.

Trent may, then, be more acceptably 'human' than Sherlock Holmes in several respects, but he has still agreed, albeit reluctantly, to take on the role of detective. He thereby knowingly renounces his fuller personality in favour of a specific set of duties and obligations. Well-versed in history and literature, he has a clear sense of the responsibilities traditionally placed upon the detective: 'I have come down in the character of avenger of blood, to hunt down the guilty, and vindicate the honour of society', he tells Cupples over a hearty breakfast. This role has, as Bentley might say, a 'heavy' mythic resonance, which Trent's joking never entirely dispels. Humanized though he may be, Trent nevertheless wears the traditional mantle of Hero, and has to go forth to do battle with monsters. Bentley's successful handling of the tale relies to a great extent on his grasp of the fact that a detective story is a kind of *romance*, that is to say, a quest in which a virtuous Hero explores a magical and dangerous territory and is prepared to put his virtues to the test in his search for the truth. The specifically sexual dimension of romance can be set aside in some versions of detective fiction, of course, but Bentley gives it a prominent place in this story. *Trent's Last Case* is, even allowing for its modern paraphernalia of motor cars, telephones, and cablegrams, the kind of romance in which a knight in shining armour might rescue some distressed damsel from a dragon. The investigation begins, indeed, with a one-eyed ogre or capitalist Cyclops, whose sinister power appears to survive even his own death. The hero then encounters other knights, not knowing yet whether they are to be friends or foes; in either case, they are greeted honourably, and with due homoerotic appreciation of their charms, as in his first meeting with John Marlowe: 'As the two approached each other, Trent noted with admiration the man's breadth of shoulder and lithe, strong figure.' A code of chivalry governs Trent's relations with men as well as with women. He greets Inspector Murch as his 'hated rival', but it soon becomes clear from their teasing banter that these two detectives are really on affectionate and respectful terms of competitive 'sportsmanship'. As Sir Gawain and other questing knights of medieval legend have to learn, however, the most serious and bewildering tests are set for them by women. In a preliminary skirmish, Trent resists easily enough the flirtatious attentions of the maid Célestine; but it is evident that greater powers of vigilance will be required in his long-delayed encounter with the newly widowed Mabel Manderson, *née* Domecq, the 'lady in black' of the seventh chapter.

As we expect, Mabel soon has Trent 'under a spell', so powerful and so traditionally cast in the formulae of medieval romance that he comes to regard her black hair as 'a forest, immeasurable, pathless and enchanted, luring him to a fatal adventure'. In the face of this mystery, Trent is shown to be disarmed, partly by an 'exaggerated chivalry' instilled in him by his mother, and partly by the erotic innocence typical of young knights. We are given to understand, by some tortuously euphemistic phrasing, that Trent is not, like Sir Galahad, a virgin, but that he is none the less dangerously inexperienced, emotionally untried. 'He went through life', we are told, 'full of a strange respect for certain feminine weakness and a very simple terror of certain feminine strength.' Mabel is, in these terms, terrifying because she is strong, in her sexuality, in her intellect, and in her judgement of herself and others. Much of what we hear about her, from her uncle Cupples and other sources, encourages us and Trent to regard her as a victim in need of chivalrous protection, but our first sight of her, through Trent's eyes, presents us with a confident 'womanhood . . . unmixed and vigorous', with a 'touch of primal joy in the excellence of [her] body'. Of greater weight is the fact that she emerges as a reasoner more consistent than Trent himself, able to belittle the detective's calculations as mere 'cleverness' beside her own deeper grasp of events. The very chivalry upon which Trent prides himself tends to blind him to possibilities that Mabel can see more easily, thus giving her the upper hand. Although certain aspects of this novel's presentation of women are doubtless predictably shallow, there is some subtlety in the unfolding of Mabel's position, and in the irony to which Trent's view of her is subjected.

Beneath the surface of the pure detective plot, then, there is a more anxious sexual plot, and both of them converge upon Mabel's boudoir: if somebody has secretly entered milady's bedchamber, then Trent as detective has to follow in the same tracks—stepping into the culprit's shoes, as it were. And in retracing this guilty route, he traces, unnervingly, the path of his own desire. His encounters with the lady in black are to be, as he accepts, a test of his own virtue, but the solution to the murder-mystery requires that Mabel's virtue too be exposed to public inspection. Aghast at the prospect of unveiling such bedroom secrets, the chivalrous detective is rendered impotent, his knowledge and his ignorance both prohibiting him from taking any further step. As in fairy-tales or romances, only a stroke of magic or grace can release him from such a trance.

The novel's sexual plot has its origin in the earlier mismatching of the young princess with the older ogre. Cupples's first, highly partisan account of his niece presents her as a young woman who had spent her life 'among people of artistic or literary propensities' but who has married into the world of high finance without recognizing 'how much soulless inhumanity that might involve'. Her five-year marriage to Sigsbee Manderson, a man more than twice her age, has made life into a 'desert' for her, since she has found that he cares only about his business. In the most impassioned speech in the book, Mabel herself explains her alienation from the moneyed world into which her rash marriage has thrown her:

Can you imagine what it must be for any one who has lived in a world where there was always creative work in the background, work with some dignity about it, men and women with professions or arts to follow, with ideals and things to believe in and quarrel about, some of them wealthy, some of them quite poor; can you think what it means to step out of that into another world where you *have* to be very rich, shamefully rich, to exist at all—where money is the only thing that counts and the first thing in everybody's thoughts—where the men who make the millions are so jaded by the work, that sport is the only thing they can occupy themselves with when they have any leisure, and the men who don't have to work are even duller than the men who do, and vicious as well; and the women live for display and silly amusements and silly immoralities; do you know how awful that life is?

Mabel's question, posed here to Philip Trent, is rhetorical, but certainly well directed, appealing as it does to Trent's own artistic values and to his hostility to the empty vulgarity of the very rich. Her ill-chosen marriage, in some ways reminiscent of Dorothea Brooke's misalliance in George Eliot's *Middlemarch*, has the effect of focusing not just the murder mystery and the potential love-interest but also the social criticisms implied by this novel. She is not just a bird in a gilded cage, but an artistically cultivated soul who recoils from the materialism of her social circle. This contrast of values is deduced by the astute detective even before he has met the widow herself: in his 'poking about' the Manderson house, he interprets the contents of the husband's library as indicating a lack of true interest in books, whereas the wife's bedroom suggests a very different inhabitant:

The room was like an unoccupied guest-chamber. Yet in every detail of furniture and decoration it spoke of an unconventional but exacting taste. Trent, as his expert eye noted the various perfection of colour and form amid which the ill-mated lady dreamed her dreams and thought her loneliest thoughts, knew that she had at least the resources of an artistic nature.

Mabel appears superior to her husband's world of urgent money-making telephone calls in that she has made herself an uncluttered space for dreaming and thinking. Perhaps more important, her taste is unconventional, suggesting an awareness of, and inner distance from, the falsehood of her position.

Murder mysteries allow us to act out, in the safety of fantasy, all manner of aggressive and vengeful wishes, beginning with the symbolic sacrifice of the principal murder-victim. In *Trent's Last Case*, the violent death of Sigsbee Manderson is, from the very start, an occasion for quiet satisfaction rather than for grief. Although this public event causes some temporary panic on the international markets, it goes conspicuously unlamented as a private loss. The first chapter invites us, if not quite to dance on Manderson's grave, then to compare the unmourned American millionaire, buried in foreign soil, with the impoverished English poet John Keats, likewise interred abroad: it is the younger and poorer man who is now richer in the posthumous respect paid by American pilgrims. From here, a continuous antagonism runs through the book, between the honourable pursuit of artistic cultivation and Manderson's ruthless and meaningless accumulation of wealth. The first comment on the murder from the artist Philip Trent is that he would not want to help hang Manderson's killer if the motive was social protest; in short, he actually condones the culling of especially brutal capitalists. His saintly, book-loving friend, Nathaniel Burton Cupples, seems to hold similar views, both on the basis of direct knowledge of Manderson and by political principle. Manderson has, notoriously, starved thirty-thousand striking Pennsylvania coal-miners and their families into submission in one of his earlier triumphs. Cupples, as a Positivist, is sympathetic to the cause of trade unions, and regards Manderson as a criminal on this account. Moreover, he is personally appalled by the millionaire's willingness to sacrifice others to his own schemes; Cupples indeed tells Manderson to his face that he is unfit to live. As a harmless old eccentric, Cupples is permitted to act as a lightning-conductor for the anti-capitalist resentment that breaks out with occasionally violent force in the novel. Elsewhere we find oblique symptoms of a struggle to the death between artist and capitalist. When Trent surveys the Manderson house, for example, he starts with the books arranged in the library:

These had a very uninspiring appearance of having been bought by the yard and never taken from their shelves. Bound with a sober luxury, the great English novelists, essayists, historians and poets stood ranged like an army struck dead in its ranks.

The striking dead of Manderson seems, in this light, almost to be an act of retribution on behalf of the imaginative writers whom he has, in some figurative sense, slaughtered. As readers, we are summoned to join the book-loving fraternity in celebrating the downfall of a Philistine monster; to cheer the David of Culture in his combat with the Goliath of Anarchy.

The contending forces in this novel then, are, on the one side, a Plutocracy of ruthless self-assertion represented by Sigsbee Manderson, and, on the other, an Aristocracy of taste represented by Trent, Cupples, Mabel Manderson, and the Oxford-educated John Marlowe. The stage is set for that peculiarly English game known to students of detective fiction as 'snobbery with violence'. The snobbery, indeed, reaches beyond the superiority of Taste to Lucre, and brings in even more troublesome elements of national conflict, principally between the English and the Americans. We start with a striking incongruity between the colossal scale and violent basis of Manderson's business empire and the small, peaceful English village of Marlstone. As Trent takes his first look at the grounds of White Gables, he reflects 'That such a place could be the scene of a crime of violence seemed fantastic; it lay so quiet and well ordered, so eloquent of disciplined service and gentle living.' Rural English tranquillity has been violated, it would seem, by bloody American gangsterism, and not for the first time in the native tradition of detective writing. Conan Doyle had often written of foreign conspiracies disrupting the peace of English villages: 'The Five Orange Pips' (1891) and 'The Dancing Men' (1903) are Sherlock Holmes stories in which violent American gangs and secret societies pursue their prey to the farthest and quietest corners of the old country. In *Trent's Last Case* itself, Marlowe draws attention to the American love of conspiratorial brotherhoods, and to the habitual resort of some sections of the American labour movement to murderous revenge. We are led, by various routes, to identify America with lawless brutality, England with disciplined, gentle living. There are, fortunately, 'decent' Americans in the story, and the existence of an American literary culture in the shape of Edgar Allan Poe and Mark Twain is acknowledged too. Otherwise, the underlying opposition goes undisturbed, except possibly by indications of a third position outside it— that of continental Europe. England may be the place of order and civility, but the true sources of artistic energy are in Italy, Germany, and France. It is to Paris that Trent must retreat in order to renew his creativity, and in Italy that Mabel waits out her term of widowhood before resuming her interest in Wagnerian opera. Mabel's assured dress-sense, too, is characterized as French rather than English or American. The novel is not, then, a simple celebration of superior Englishness, and the fact that a New York publisher was the first to accept it seems to underline this qualification. There are hints, mercifully slight, of an anti-semitic prejudice that was particularly virulent in the circles around Bentley's friend G. K. Chesterton, but Bentley does not make his sinister plutocrat a Jew as many British writers of this period might have done. Despite the Manderson name, which seems to suggest Jewish descent, the financier turns out to be the shamefaced carrier of 'mixed blood' deriving ultimately from the eighteenth-century Iroquois chieftain Montour and his French wife—a genealogy that sends us back to the mythic America of Fenimore Cooper's *The Last of the Mohicans*.

The element of *Trent's Last Case* that sits most uncomfortably with English self-satisfaction is its complete lack of trust in the Law. Of course, most detective fictions assume that the unimaginative methods of the standard police investigation will often result in the guilty escaping justice while the innocent fall under suspicion; but in the end the superior perceptions of the detective hero or heroine will sort these errors out, leaving the ever-reliable courts to mete out appropriate punishment. In Bentley's novel, however, the legal system is denied even this bare endorsement. The coroner, for instance, is shown to guide the jury improperly towards a particular explanation of Manderson's death. When they come to discuss the criminal courts, Trent and Cupples are both convinced that judge and jury offer little protection to innocent people accused of murder or other hanging offences on circumstantial evidence. Trent goes further, and charges the British imperial administration with running police states and encouraging miscarriages of justice:

As for attempts being made by malevolent persons to fix crimes upon innocent men, of course it is constantly happening. It's a marked feature, for instance, of all systems of rule by coercion, whether in Ireland or Russia or India or Korea; if the police cannot get hold of a man they think dangerous by fair means, they do it by foul.

No self-respecting detective could, in these circumstances, merely 'hand over' a suspect to the Law without himself becoming embroiled in a possible hanging of the innocent. The legal system is susceptible to error and to improper manipulation, and so is the mind of the professional or amateur detective. As soon as hypotheses of guilt start to be formed, nobody is safe, the novel seems to imply in its more sombre moments.

There is, though, a lighter side to the book, at least when Trent is able to turn aside momentarily from the worrying implications of the Manderson murder. In his banter with Cupples and others, the detective hero represents an effervescent re-assertion of youthful life against the dead hand of the past, represented by the late Sigsbee Manderson. His festive nonsense and laughter are not just diverting changes of mood but substantial values in themselves. Crucially, they allow Trent to release himself from the detective's burden of infallible rationality, to mock his previous fixed ideas, and to accept new surprises.

Philip Trent's merriment often takes a literary turn. As we have noticed, he likes to invoke the poets, although usually upon unpoetical occasions and from faulty memory. For most of those modern readers who have not been brought up on Scott and Tennyson, this may be a perplexing mannerism, and for some who half-remember the detective's fleeting allusions, it can be maddening. I have not been able to provide for this edition an apparatus of explanatory notes, but it may be helpful to indicate here some of the sources of Trent's verse quotations. Virtually all the poetical allusions are to the Romantic and post-Romantic poets of the nineteenth century. Rather more variable is the degree of inaccuracy or of ludicrous travesty to which their lines are subjected in quotation. Trent's first playful allusion is, suitably enough in the light of his own semi-chivalric status, to an Arthurian poem, *The Passing of Arthur*, from Tennyson's *Idylls of the King*. This he pays the compliment of quoting accurately ('I am blown along a wandering wind, and hollow, hollow, hollow and delight'). His next flourish, however, in the same conversation with the editor of the *Record*, involves adapting Ralph Waldo Emerson's mystical poem 'Brahma' so that its lines

> I am the doubter and the doubt
> And I the hymn the Brahmin sings

can serve as an expression of enthusiasm for railways, in Trent's declaration that 'I am the stoker and the stoked. I am the song the porter sings.' Sir James Molloy, on the other end of the telephone line, is, not surprisingly, a little baffled. In his later resort to the poets, Trent varies between comic travesty and respectful echo. Thus Byron's lines from the third canto of *Don Juan*,

> Fill high the cup with Samian wine!
> Leave battles to the Turkish hordes,

are amended so as to substitute 'soda' for 'battles'; and Wordsworth's tribute to the Child in his Immortality Ode—

> Thou, whose exterior semblance doth belie
> Thy Soul's immensity

—is garbled as 'Your outward semblance doth belie your soul's immensity'. Some verse quotations are merely curtailed or slightly altered to fit the immediate context, surviving relatively unscathed: 'May heaven our simple lives prevent from luxury's contagion, weak and vile' comes from Robert Burns's 'The Cotter's Saturday Night', while 'From childhood's hour I've seen my fondest hopes decay' is quoted from Tom Moore's poem 'The Fire-Worshippers'; and the warning 'For want of you the world's course will not fail' is adapted from Coventry Patmore's *Magna est Veritas*. More accurately reproduced are 'The dun deer's hide on fleeter foot was never tied', from Walter Scott's *The Lady of the Lake*, 'Cease! Drain not to its dregs the urn of bitter prophecy', from Shelley's *Hellas*, 'Bring me a tablet whiter than a star, or hand of hymning angel', from Keats's sonnet

'On Leaving Some Friends at an Early Hour', and almost an entire quatrain—'A star upon your birthday burned, whose fierce, serene, red, pulseless planet never yearned in heaven'—from A. C. Swinburne's 'Faustine'.

The variety, and the spontaneous impertinence, of these and other allusions in Trent's conversation are intended to reinforce the central distinction between this artist-detective and such a sleuth as Sherlock Holmes: whereas Holmes has no real interests outside his consuming obsession with crime, Philip Trent has an imaginative life extending through other realms. He is, accordingly, able to step back from the claims of investigative deduction: after all, there are more important things. Bentley succeeded in making the detective story more humane by taking it more lightly, placing at its centre the nonchalance of the English gentleman-amateur. This figure would, in turn, ossify into a cliché of British detective writing, but in *Trent's Last Case* we can witness his arrival in all its confident freshness.

SELECT BIBLIOGRAPHY

EDITIONS

The Woman in Black (New York: Century, 1913).
Trent's Last Case (London: Nelson, 1913).
Trent's Last Case (New York: Knopf, 1929).
Trent's Last Case (Harmondsworth: Penguin Books, 1937).

OTHER WORKS BY BENTLEY

Biography for Beginners (as 'E. Clerihew') (London: Werner Laurie, 1905).
More Biography (London: Methuen, 1929).
Trent's Own Case, with H. Warner Allen (London: Constable, 1936).
Trent Intervenes (London: Nelson, 1938).
Baseless Biography (London: Constable, 1939).
Those Days (London: Constable, 1940).
Elephant's Work (London: Hodder & Stoughton, 1950).
Clerihews Complete (London: Werner Laurie, 1951).
The First Clerihews (Oxford: Oxford University Press, 1982).

BIOGRAPHY AND CRITICISM

Bentley, Nicolas, *A Version of the Truth* (London: Deutsch, 1960).
Binyon, T. J., *Murder Will Out: The Detective in Fiction* (Oxford: Oxford University Press, 1989).
Haycraft, Howard, *Murder for Pleasure: The Life and Times of the Detective Story* (New York: Appleton, 1941).
Leitch, Thomas M., 'E. C. Bentley', in Benstock, Bernard, and Thomas F. Staley (eds.), *Dictionary of Literary Biography, Volume 70: British Mystery Writers, 1860–1919* (Detriot: Gale, 1988), 23–9.
Menon, K. R., *A Guide to E. C. Bentley's 'Trent's Last Case'* (Singapore: India Publishing House, 1957).
Panek, LeRoy, *Watteau's Shepherds: The Detective Novel in Britain, 1914–1940* (Bowling Green, Ohio: Bowling Green University Press, 1979).
Sayers, Dorothy L., Introduction to E. C. Bentley, *Trent's Last Case* (New York: Perennial Library, 1978), pp. x–xiii.
Thomson, H. Douglas, *Masters of Mystery: A Study of the Detective Story* (London: Collins, 1931).

Trent's Last Case

TO

GILBERT KEITH CHESTERTON

My dear Gilbert,

I dedicate this story to you. First: because the only really noble motive I had in writing it was the hope that you would enjoy it. Second: because I owe you a book in return for 'The Man Who Was Thursday'. Third: because I said I would when I unfolded the plan of it to you, surrounded by Frenchmen, two years ago. Fourth: because I remember the past.

I have been thinking again today of those astonishing times when neither of us ever looked at a newspaper; when we were purely happy in the boundless consumption of paper, pencils, tea, and our elders' patience; when we embraced the most severe literature, and ourselves produced such light reading as was necessary; when (in the words of Canada's poet) we studied the works of nature, also those little frogs; when, in short, we were extremely young.

For the sake of that age I offer you this book.

Yours always,
E. C. BENTLEY

CHAPTER I

Bad News

BETWEEN what matters and what seems to matter, how should the world we know judge wisely?

When the scheming, indomitable brain of Sigsbee Manderson was scattered by a shot from an unknown hand, that world lost nothing worth a single tear; it gained something memorable in a harsh reminder of the vanity of such wealth as this dead man had piled up—without making one loyal friend to mourn him, without doing an act that could help his memory to the least honour. But when the news of his end came, it seemed to those living in the great vortices of business as if the earth too shuddered under a blow.

In all the lurid commercial history of his country there had been no figure that had so imposed itself upon the mind of the trading world. He had a niche apart in its temples. Financial giants, strong to direct and augment the forces of capital, and taking an approved toll in millions for their labour, had existed before; but in the case of Manderson there had been this singularity, that a pale halo of piratical romance, a thing especially dear to the hearts of his countrymen, had remained incongruously about his head through the years when he stood in every eye as the unquestioned guardian of stability, the stamper-out of manipulated crises, the foe of the raiding chieftains that infest the borders of Wall Street.

The fortune left by his grandfather, who had been one of those chieftains on the smaller scale of his day, had descended to him with accretion through his father, who during a long life had quietly continued to lend money and never had margined a stock. Manderson, who had at no time known what it was to be without large sums to his hand, should have been altogether of that newer American plutocracy which is steadied by the tradition and habit of great wealth. But it was not so. While his nurture and education had taught him European ideas of a rich man's proper external circumstance; while they had rooted in him an instinct for quiet magnificence, the larger costliness which does not shriek of itself with a thousand tongues; there had been handed on to him nevertheless much of the Forty-Niner and financial buccaneer, his forbear. During that first period of his business career which had been called his early bad manner, he had been little more than a gambler of genius, his hand against every man's— an infant prodigy who brought to the enthralling pursuit of speculation a brain better endowed than any opposed to it. At St Helena it was laid down that war is *une belle occupation*; and so the young Manderson had found the multitudinous and complicated dog-fight of the Stock Exchange of New York.

Then came his change. At his father's death, when Manderson was thirty years old, some new revelation of the power and the glory of the god he served seemed to have come upon him. With the sudden, elastic adaptability of his nation he turned to steady labour in his father's banking business, closing his ears to the sound of the battles of the Street. In a few years he came to control all the activity of the great firm whose unimpeached conservatism, safety, and financial weight lifted it like a cliff above the angry sea of the markets. All mistrust founded on the performances of his youth had vanished. He was quite plainly a different man. How the change came about none could with authority say, but there was a story of certain last words spoken by his father, whom alone he had respected and perhaps loved.

He began to tower above the financial situation. Soon his name was current in the bourses of the world. One who spoke the name of Manderson called up a vision of all that was broad-based and firm in the vast wealth of the United States. He planned great combinations of capital, drew together and centralized industries of continental scope, financed with unerring judgement the large designs of state or of private enterprise. Many a time when he 'took hold' to smash a strike, or to federate the ownership of some great field of labour, he sent ruin upon a multitude of tiny homes; and if miners or steel-workers or cattlemen defied him and invoked disorder, he could be more lawless and ruthless than they. But this was done in the pursuit of legitimate business ends. Tens of thousands of the poor might curse his name, but the financier and the speculator execrated him no more. He stretched a hand to protect or to manipulate the power of wealth in every corner of the country. Forcible, cold, and unerring, in all he did he ministered to the national lust for magnitude; and a grateful country surnamed him the Colossus.

But there was an aspect of Manderson in this later period that lay long unknown and unsuspected save by a few, his secretaries and lieutenants and certain of the associates of his bygone hurling time. This little circle knew that Manderson, the pillar of sound business and stability in the markets, had his hours of nostalgia for the lively times when the Street had trembled at his name. It was, said one of them, as if Blackbeard had settled down as a decent merchant in Bristol on the spoils of the Main. Now and then the pirate would glare suddenly out, the knife in his teeth and the sulphur matches sputtering in his hatband. During such spasms of reversion to type a score of tempestuous raids upon the market had been planned on paper in the inner room of the offices of Manderson, Colefax and Company. But they were never carried out. Blackbeard would quell the mutiny of his old self within him and go soberly down to his counting-house—humming a stave or two of 'Spanish Ladies', perhaps, under his breath. Manderson would allow himself the harmless satisfaction, as soon as the time for action had gone by, of pointing out to some Rupert of the markets how a *coup* worth a million to the depredator might have been made. 'Seems to me,' he would say almost wistfully, 'the Street is getting to be a mighty dull place since I quit.' By slow degrees this amiable weakness of the Colossus became known to the business world, which exulted greatly in the knowledge.

At the news of his death panic went through the markets like a hurricane; for it came at a luckless time. Prices tottered and crashed like towers in an earthquake. For two days Wall Street was a clamorous inferno of pale despair. All over the United States, wherever speculation had its devotees, went a waft of ruin, a plague of suicide. In Europe also not a few took with their own hands lives that had become pitiably linked to the destiny of a financier whom most of them had never seen. In Paris a well-known banker walked quietly out of the Bourse and fell dead upon the broad steps among the raving crowd of Jews, a phial crushed in his hand. In Frankfort one leapt from the Cathedral top, leaving a redder stain where he struck the red tower. Men stabbed and shot and strangled themselves, drank death or breathed it as the air, because in a lonely corner of England the life had departed from one cold heart vowed to the service of greed.

The blow could not have fallen at a more disastrous moment. It came when Wall Street was in a condition of suppressed 'scare'— suppressed, because for a week past the great interests known to act with or to be actually controlled by the Colossus had been desperately combating the effects of the sudden arrest of Lucas Hahn, and the exposure of his plundering of the Hahn banks. This bombshell, in its turn, had fallen at a time when the market had been 'boosted' beyond its real strength. In the language of the place, a slump was due. Reports from the corn-lands had not been good, and there had been two or three railway statements which had been expected to be much better than they were. But at what-ever point in the vast area of speculation the shudder of the threat-ened break had been felt, 'the Manderson crowd' had stepped in and held the market up. All through the week the speculator's mind, as shallow as it is quick-witted, as sentimental as greedy, had seen in this the hand of the giant stretched out in protection from afar. Manderson, said the newspapers in chorus, was in hourly communica-tion with his lieutenants in the Street. One journal was able to give in round figures the sum spent on cabling between New York and Marlstone in the past twenty-four hours; it told how a small staff of expert operators had been sent down by the Post Office authorities to Marlstone to deal with the flood of messages. Another revealed that Manderson, on the first news of the Hahn crash, had arranged to abandon his holiday and return home by the *Lusitania*; but that he soon had the situation so well in hand that he had determined to remain where he was.

All this was falsehood, more or less consciously elaborated by the 'finance editors', consciously initiated and encouraged by the shrewd business men of the Manderson group, who knew that nothing could better help their plans than this illusion of hero-worship—knew also that no word had come from Manderson in answer to their messages, and that Howard B. Jeffrey, of Steel and Iron fame, was the true organizer of victory. So they fought down apprehension through four feverish days, and minds grew calmer. On Saturday, though the ground beneath the feet of Mr Jeffrey yet rumbled now and then with Etna-mutterings of disquiet, he deemed his task almost done. The market was firm, and slowly advancing. Wall Street turned to its sleep of Sunday, worn out but thankfully at peace.

In the first trading hour of Monday a hideous rumour flew round the sixty acres of the financial district. It came into being as the lightning comes—a blink that seems to begin nowhere; though it is to be suspected that it was first whispered over the telephone—together with an urgent selling order—by some employee in the cable service. A sharp spasm convulsed the convalescent share-list. In five minutes the dull noise of the kerbstone market in Broad Street had leapt to a high note of frantic interrogation. From within the hive of the Exchange itself could be heard a droning hubbub of fear, and men rushed hatless in and out. Was it true? asked every man; and every man replied, with trembling lips, that it was a lie put out by some unscrupulous 'short' interest seeking to cover itself. In another quarter of an hour news came of a sudden and ruinous collapse of 'Yankees' in London at the close of the Stock Exchange day. It was enough. New York had still four hours' trading in front of her. The strategy of pointing to Manderson as the saviour and warden of the markets had recoiled upon its authors with annihilating force, and Jeffrey, his ear at his private telephone, listened to the tale of disaster with a set jaw. The new Napoleon had lost his Marengo. He saw the whole financial landscape sliding and falling into chaos before him. In half an hour the news of the finding of Manderson's body, with the inevitable rumour that it was suicide, was printing in a dozen newspaper offices; but before a copy reached Wall Street the tornado of the panic was in full fury, and Howard B. Jeffrey and his collaborators were whirled away like leaves before its breath.

All this sprang out of nothing.

Nothing in the texture of the general life had changed. The corn had not ceased to ripen in the sun. The rivers bore their barges and gave power to a myriad engines. The flocks fattened on the pastures, the herds were unnumbered. Men laboured everywhere in the various servitudes to which they were born, and chafed not more than usual in their bonds. Bellona tossed and murmured as ever, yet still slept her uneasy sleep. To all mankind save a million or two of half-crazed gamblers, blind to all reality, the death of Manderson meant nothing; the life and work of the world went on. Weeks before he died strong hands had been in control of every wire in the huge network of commerce and industry that he had supervised. Before his corpse was buried his countrymen had made a strange discovery—that the existence of the potent engine of monopoly that went by the name of Sigsbee Manderson had not been a condition of even material prosperity. The panic blew itself out in two days, the pieces were picked up, the bankrupts withdrew out of sight; the market 'recovered a normal tone'.

While the brief delirium was yet subsiding there broke out a domestic scandal in England that suddenly fixed the attention of two continents. Next morning the Chicago Limited was wrecked, and the same day a notable politician was shot down in cold blood by his wife's brother in the streets of New Orleans. Within a week of its rising, 'the Manderson story', to the trained sense of editors through-out the Union, was 'cold'. The tide of American visitors pouring through Europe made eddies round the memorial or statue of many a man who had died in poverty; and never thought of their most famous plutocrat. Like the poet who died in Rome, so young and poor, a hundred years ago, he was buried far away from his own land; but for all the men and women of Manderson's people who flock round the tomb of Keats in the cemetery under the Monte Testaccio, there is not one, nor ever will be, to stand in reverence by the rich man's grave beside the little church of Marlstone.

CHAPTER II

Knocking the Town Endways

IN the only comfortably furnished room in the offices of the *Record*, the telephone on Sir James Molloy's table buzzed. Sir James made a motion with his pen, and Mr Silver, his secretary, left his work and came over to the instrument.

'Who is that?' he said. 'Who? . . . I can't hear you. . . . Oh, it's Mr Bunner, is it? . . . Yes, but . . . I know, but he's fearfully busy this afternoon. Can't you . . . Oh, really? Well, in that case—just hold on, will you?'

He placed the receiver before Sir James. 'It's Calvin Bunner, Sigsbee Manderson's right-hand man,' he said concisely. 'He insists on speaking to you personally. Says it is the gravest piece of news. He is talking from the house down by Bishopsbridge, so it will be necessary to speak clearly.'

Sir James looked at the telephone, not affectionately, and took up the receiver. 'Well?' he said in his strong voice, and listened. 'Yes,' he said. The next moment Mr Silver, eagerly watching him, saw a look of amazement and horror. 'Good God!' murmured Sir James. Clutching the instrument, he slowly rose to his feet, still bending ear intently. At intervals he repeated 'Yes.' Presently, as he listened, he glanced at the clock, and spoke quickly to Mr Silver over the top of the transmitter. 'Go and hunt up Figgis and young Williams. Hurry.' Mr Silver darted from the room.

The great journalist was a tall, strong, clever Irishman of fifty, swart and black-moustached, a man of untiring business energy, well known in the world, which he understood very thoroughly, and played upon with the half-cynical competence of his race. Yet was he without a touch of the charlatan: he made no mysteries, and no pretences of knowledge, and he saw instantly through these in others. In his handsome, well-bred, well-dressed appearance there was something a little sinister when anger or intense occupation put its imprint about his eyes and brow; but when his generous nature was under no restraint he was the most cordial of men. He was managing director of the company which owned that most powerful morning paper, the *Record*, and also that most indispensable evening paper, the *Sun*, which had its offices on the other side of the street. He was, moreover, editor-in-chief of the *Record*, to which he had in the course of years attached the most variously capable personnel in the country. It was a maxim of his that where you could not get gifts, you must do the best you could with solid merit; and he employed a great deal of both. He was respected by his staff as few are respected in a profession not favourable to the growth of the sentiment of reverence.

'You're sure that's all?' asked Sir James, after a few minutes of earnest listening and questioning. 'And how long has this been known? . . . Yes, of course, the police are; but the servants? Surely it's all over the place down there by now. . . . Well, we'll have a try. . . . Look here, Bunner, I'm infinitely obliged to you about this. I owe you a good turn. You know I mean what I say. Come and see me the first day you get to town. . . . All right, that's understood. Now I must act on your news. Goodbye.'

Sir James hung up the receiver, and seized a railway timetable from the rack before him. After a rapid consultation of this oracle, he flung it down with a forcible word as Mr Silver hurried into the room, followed by a hard-featured man with spectacles, and a youth with an alert eye.

'I want you to jot down some facts, Figgis,' said Sir James, banishing all signs of agitation and speaking with a rapid calmness. 'When you have them, put them into shape just as quick as you can for a special edition of the *Sun*.' The hard-featured man nodded and glanced at the clock, which pointed to a few minutes past three; he pulled out a notebook and drew a chair up to the big writing-table. 'Silver,' Sir James went on, 'go and tell Jones to wire our local correspondent very urgently, to drop everything and get down to Marlstone at once. He is not to say why in the telegram. There must not be an unnecessary word about this news until the *Sun* is on the streets with it—you all understand. Williams, cut across the way and tell Mr Anthony to hold himself ready for a two-column opening that will knock the town endways. Just tell him that he must take all measures and precautions for a scoop. Say that Figgis will be over in five minutes with the facts, and that he had better let him write up the story in his private room. As you go, ask Miss Morgan to see me here at once, and tell the telephone people to see if they can get Mr Trent on the wire for me. After seeing Mr Anthony, return here and stand by.' The alert-eyed young man vanished like a spirit.

Sir James turned instantly to Mr Figgis, whose pencil was poised over the paper. 'Sigsbee Manderson has been murdered,' he began quickly and clearly, pacing the floor with his hands behind him. Mr Figgis scratched down a line of shorthand with as much emotion as if he had been told that the day was fine—the pose of his craft. 'He and his wife and two secretaries have been for the past fortnight at the house called White Gables, at Marlstone, near Bishopsbridge. He bought it four years ago. He and Mrs Manderson have since spent a part of each summer there. Last night he went to bed about half-past eleven, just as usual. No one knows when he got up and left the house. He was not missed until this morning. About ten o'clock his body was found by a gardener. It was lying by a shed in the grounds. He was shot in the head, through the left eye. Death must have been instantaneous. The body was not robbed, but there were marks on the wrists which pointed to a struggle having taken place. Dr Stock, of Marlstone, was at once sent for, and will conduct the post-mortem examination. The police from Bishopsbridge, who were soon on the spot, are reticent, but it is believed that they are quite without a clue to the identity of the murderer. There you are, Figgis. Mr Anthony is expecting you. Now I must telephone him and arrange things.'

Mr Figgis looked up. 'One of the ablest detectives at Scotland Yard,' he suggested, 'has been put in charge of the case. It's a safe statement.'

'If you like,' said Sir James.

'And Mrs Manderson? Was she there?'

'Yes. What about her?'

'Prostrated by the shock,' hinted the reporter, 'and sees nobody. Human interest.'

'I wouldn't put that in, Mr Figgis,' said a quiet voice. It belonged to Miss Morgan, a pale, graceful woman, who had silently made her appearance while the dictation was going on. 'I have seen Mrs Manderson,' she proceeded, turning to Sir James. 'She looks quite

healthy and intelligent. Has her husband been murdered? I don't
think the shock would prostrate her. She is more likely to be doing
all she can to help the police.'

'Something in your own style, then, Miss Morgan,' he said with a
momentary smile. Her imperturbable efficiency was an office proverb. 'Cut it out, Figgis. Off you go! Now, madam, I expect you know
what I want.'

'Our Manderson biography happens to be well up to date,' replied
Miss Morgan, drooping her dark eyelashes as she considered
the position. 'I was looking over it only a few months ago. It is practically ready for tomorrow's paper. I should think the *Sun* had better
use the sketch of his life they had about two years ago, when he went
to Berlin and settled the potash difficulty. I remember it was a very
good sketch, and they won't be able to carry much more than that.
As for our paper, of course we have a great quantity of cuttings, mostly
rubbish. The sub-editors shall have them as soon as they come
in. Then we have two very good portraits that are our own property; the best is a drawing Mr Trent made when they were both
on the same ship somewhere. It is better than any of the photographs;
but you say the public prefers a bad photograph to a good drawing.
I will send them down to you at once, and you can choose. As far
as I can see, the *Record* is well ahead of the situation, except that you
will not be able to get a special man down there in time to be of
any use for tomorrow's paper.'

Sir James sighed deeply. 'What are we good for, anyhow?'
he enquired dejectedly of Mr Silver, who had returned to his desk.
'She even knows Bradshaw by heart.'

Miss Morgan adjusted her cuffs with an air of patience. 'Is there
anything else?' she asked, as the telephone bell rang.

'Yes, one thing,' replied Sir James, as he took up the receiver.
'I want you to make a bad mistake some time, Miss Morgan—
an everlasting bloomer—just to put us in countenance.' She permitted herself the fraction of what would have been a charming smile as
she went out.

'Anthony?' asked Sir James, and was at once deep in consultation
with the editor on the other side of the road. He seldom entered
the *Sun* building in person; the atmosphere of an evening paper,
he would say, was all very well if you liked that kind of thing.
Mr Anthony, the Murat of Fleet Street, who delighted in riding

the whirlwind and fighting a tumultuous battle against time, would say the same of a morning paper.

It was some five minutes later that a uniformed boy came in to say that Mr Trent was on the wire. Sir James abruptly closed his talk with Mr Anthony.

'They can put him through at once,' he said to the boy.

'Hullo!' he cried into the telephone after a few moments.

A voice in the instrument replied, 'Hullo be blowed! What do you want?'

'This is Molloy,' said Sir James.

'I know it is,' the voice said. 'This is Trent. He is in the middle of painting a picture, and he has been interrupted at a critical moment. Well, I hope it's something important, that's all!'

'Trent,' said Sir James impressively, 'it is important. I want you to do some work for us.'

'Some play, you mean,' replied the voice. 'Believe me, I don't want a holiday. The working fit is very strong. I am doing some really decent things. Why can't you leave a man alone?'

'Something very serious has happened.'

'What?'

'Sigsbee Manderson has been murdered—shot through the brain—and they don't know who has done it. They found the body this morning. It happened at his place near Bishopsbridge.' Sir James proceeded to tell his hearer, briefly and clearly, the facts that he had communicated to Mr Figgis. 'What do you think of it?' he ended.

A considering grunt was the only answer.

'Come now,' urged Sir James.

'Tempter!'

'You will go down?'

There was a brief pause.

'Are you there?' said Sir James.

'Look here, Molloy,' the voice broke out querulously, 'the thing may be a case for me, or it may not. We can't possibly tell. It may be a mystery; it may be as simple as bread and cheese. The body not being robbed looks interesting, but he may have been outed by some wretched tramp whom he found sleeping in the grounds and tried to kick out. It's the sort of thing he would do. Such a murderer might easily have sense enough to know that to leave the money and valuables was the safest thing. I tell you frankly, I wouldn't have a hand in hanging a poor devil who had let daylight into a man like Sig Manderson as a measure of social protest.'

Sir James smiled at the telephone—a smile of success. 'Come, my boy, you're getting feeble. Admit you want to go and have a look at the case. You know you do. If it's anything you don't want to handle, you're free to drop it. By the by, where are you?'

'I am blown along a wandering wind,' replied the voice irresolutely, 'and hollow, hollow, hollow all delight.'

'Can you get here within an hour?' persisted Sir James.

'I suppose I can,' the voice grumbled. 'How much time have I?'

'Good man! Well, there's time enough—that's just the worst of it. I've got to depend on our local correspondent for tonight. The only good train of the day went half an hour ago. The next is a slow one, leaving Paddington at midnight. You could have the Buster, if you like'—Sir James referred to a very fast motor car of his—'but you wouldn't get down in time to do anything tonight.'

'And I'd miss my sleep. No, thanks. The train for me. I am quite fond of railway travelling, you know; I have a gift for it. I am the stoker and the stoked. I am the song the porter sings.'

'What's that you say?'

'It doesn't matter,' said the voice sadly. 'I say,' it continued, 'will your people look out a hotel near the scene of action, and telegraph for a room?'

'At once,' said Sir James. 'Come here as soon as you can.'

He replaced the receiver. As he turned to his papers again a shrill outcry burst forth in the street below. He walked to the open window. A band of excited boys was rushing down the steps of the *Sun* building and up the narrow thoroughfare toward Fleet Street. Each carried a bundle of newspapers and a large broadsheet with the simple legend:

> # MURDER
> # OF
> # SIGSBEE
> # MANDERSON

Sir James smiled and rattled the money in his pockets cheerfully.

'It makes a good bill,' he observed to Mr Silver, who stood at his elbow.

Such was Manderson's epitaph.

CHAPTER III

Breakfast

AT about eight o'clock in the morning of the following day Mr Nathaniel Burton Cupples stood on the veranda of the hotel at Marlstone. He was thinking about breakfast. In his case the colloquialism must be taken literally: he really was thinking about breakfast, as he thought about every conscious act of his life when time allowed deliberation. He reflected that on the preceding day the excitement and activity following upon the discovery of the dead man had disorganized his appetite, and led to his taking considerably less nourishment than usual. This morning he was very hungry, having already been up and about for an hour; and he decided to allow himself a third piece of toast and an additional egg; the rest as usual. The remaining deficit must be made up at luncheon, but that could be gone into later.

So much being determined, Mr Cupples applied himself to the enjoyment of the view for a few minutes before ordering his meal. With a connoisseur's eye he explored the beauty of the rugged coast, where a great pierced rock rose from a glassy sea, and the ordered loveliness of the vast tilted levels of pasture and tillage and woodland that sloped gently up from the cliffs toward the distant moor. Mr Cupples delighted in landscape.

He was a man of middle height and spare figure, nearly sixty years old, by constitution rather delicate in health, but wiry and active for his age. A sparse and straggling beard and moustache did not conceal a thin but kindly mouth; his eyes were keen and pleasant; his sharp nose and narrow jaw gave him very much of a clerical air, and this impression was helped by his commonplace dark clothes and soft black hat. The whole effect of him, indeed, was priestly. He was a man of unusually conscientious, industrious, and orderly mind, with little imagination. His father's household had been used to recruit its domestic establishment by means of advertisements in which it was truthfully described as a serious family. From that fortress of gloom he had escaped with two saintly gifts somehow unspoiled: an inexhaustible kindness of heart, and a capacity for innocent gaiety which owed nothing to humour. In an earlier day and with a clerical training he might have risen to the scarlet hat. He was, in fact, a highly regarded member of the London Positivist Society, a retired banker, a widower without children. His austere but not unhappy life was spent largely among books and in museums; his profound and patiently accumulated knowledge of a number of curiously disconnected subjects which had stirred his interest at different times had given him a place in the quiet, half-lit world of professors and curators and devotees of research; at their amiable, unconvivial dinner parties he was most himself. His favourite author was Montaigne.

Just as Mr Cupples was finishing his meal at a little table on the veranda, a big motor car turned into the drive before the hotel. 'Who is this?' he enquired of the waiter. 'Id is der manager,' said the young man listlessly. 'He have been to meed a gendleman by der train.'

The car drew up and the porter hurried from the entrance. Mr Cupples uttered an exclamation of pleasure as a long, loosely built man, much younger than himself, stepped from the car and mounted the veranda, flinging his hat on a chair. His high-boned, quixotic face wore a pleasant smile; his rough tweed clothes, his hair and short moustache were tolerably untidy.

'Cupples, by all that's miraculous!' cried the man, pouncing upon Mr Cupples before he could rise, and seizing his outstretched hand in a hard grip. 'My luck is serving me today,' the newcomer went on spasmodically. 'This is the second slice within an hour. How are you, my best of friends? And why are you here? Why sit'st thou by that ruined breakfast? Dost thou its former pride recall, or ponder how it passed away? I *am* glad to see you!'

'I was half expecting you, Trent,' Mr Cupples replied, his face wreathed in smiles. 'You are looking splendid, my dear fellow. I will tell you all about it. But you cannot have had your own breakfast yet. Will you have it at my table here?'

'Rather!' said the man. 'An enormous great breakfast, too—with refined conversation and tears of recognition never dry. Will you get young Siegfried to lay a place for me while I go and wash? I shan't be three minutes.' He disappeared into the hotel, and Mr Cupples, after a moment's thought, went to the telephone in the porter's office.

He returned to find his friend already seated, pouring out tea, and showing an unaffected interest in the choice of food. 'I expect this to be a hard day for me,' he said, with the curious jerky utterance which seemed to be his habit. 'I shan't eat again till the evening, very likely. You guess why I'm here, don't you?'

'Undoubtedly,' said Mr Cupples. 'You have come down to write about the murder.'

'That is rather a colourless way of stating it,' the man called Trent replied, as he dissected a sole. 'I should prefer to put it that I have come down in the character of avenger of blood, to hunt down the guilty, and vindicate the honour of society. That is my line of business. Families waited on at their private residences. I say, Cupples, I have made a good beginning already. Wait a bit, and I'll tell you.' There was a silence, during which the newcomer ate swiftly and abstractedly, while Mr Cupples looked on happily.

'Your manager here,' said the tall man at last, 'is a fellow of remarkable judgement. He is an admirer of mine. He knows more about my best cases than I do myself. The *Record* wired last night to say I was coming, and when I got out of the train at seven o'clock this morning, there he was waiting for me with a motor car the size of a haystack. He is beside himself with joy at having me here. It is fame.' He drank a cup of tea and continued: 'Almost his first words were to ask me if I would like to see the body of the murdered man—if so, he thought he could manage it for me. He is as keen as a razor. The body lies in Dr Stock's surgery, you know, down in the village, exactly as it was when found. It's to be post-mortem'd this morning, by the way, so I was only just in time. Well, he ran me down here to the doctor's, giving me full particulars about the case all the way. I was pretty well *au fait* by the time we arrived. I suppose the manager of a place like this has some sort of a pull with the doctor. Anyhow, he made no difficulties, nor did the constable on duty, though he was careful to insist on my not giving him away in the paper.'

'I saw the body before it was removed,' remarked Mr Cupples. 'I should not have said there was anything remarkable about it, except that the shot in the eye had scarcely disfigured the face at all, and caused scarcely any effusion of blood, apparently. The wrists were scratched and bruised. I expect that, with your trained faculties, you were able to remark other details of a suggestive nature.'

'Other details, certainly; but I don't know that they suggest anything. They are merely odd. Take the wrists, for instance. How was it you could see bruises and scratches on them? I dare say you saw something of Manderson down here before the murder.'

'Certainly,' Mr Cupples said.

'Well, did you ever see his wrists?'

Mr Cupples reflected. 'No. Now you raise the point, I am reminded that when I interviewed Manderson here he was wearing stiff cuffs, coming well down over his hands.'

'He always did,' said Trent. 'My friend the manager says so. I pointed out to him the fact you didn't observe, that there were no cuffs visible, and that they had, indeed, been dragged up inside the coat-sleeves, as yours would be if you hurried into a coat without pulling your cuffs down. That was why you saw his wrists.'

'Well, I call that suggestive,' observed Mr Cupples mildly. 'You might infer, perhaps, that when he got up he hurried over his dressing.'

'Yes, but did he? The manager said just what you say. "He was always a bit of a swell in his dress," he told me, and he drew the inference that when Manderson got up in that mysterious way, before the house was stirring, and went out into the grounds, he was in a great hurry. "Look at his shoes," he said to me: "Mr Manderson was always specially neat about his footwear. But those shoe-laces were tied in a hurry." I agreed. "And he left his false teeth in his room," said the manager. "Doesn't *that* prove he was flustered and hurried?" I allowed that it looked like it. But I said, "Look here: if he was so very much pressed, why did he part his hair so carefully? That parting is a work of art. Why did he put on so much? for he had on a complete outfit of underclothing, studs in his shirt, sock-suspenders, a watch and chain, money and keys and things in his pockets." That's what I said to the manager. He couldn't find an explanation. Can you?"

Mr Cupples considered. 'Those facts might suggest that he was hurried only at the end of his dressing. Coat and shoes would come last.'

'But not false teeth. You ask anybody who wears them. And besides, I'm told he hadn't washed at all on getting up, which in a neat man looks like his being in a violent hurry from the beginning. And here's another thing. One of his waistcoat pockets was lined with wash-leather for the reception of his gold watch. But he had put his watch into the pocket on the other side. Anybody who has settled habits can see how odd that is. The fact is, there are signs of great agitation and haste, and there are signs of exactly the opposite. For the present I am not guessing. I must reconnoitre the ground first, if I can manage to get the right side of the people of the house.' Trent applied himself again to his breakfast.

Mr Cupples smiled at him benevolently. 'That is precisely the point,' he said, 'on which I can be of some assistance to you.' Trent glanced up in surprise. 'I told you I half expected you. I will explain the situation. Mrs Manderson, who is my niece—'

'What!' Trent laid down his knife and fork with a clash. 'Cupples, you are jesting with me.'

'I am perfectly serious, Trent, really,' returned Mr Cupples earnestly. 'Her father, John Peter Domecq, was my wife's brother. I never mentioned my niece or her marriage to you before, I suppose. To tell the truth, it has always been a painful subject to me, and I have avoided discussing it with anybody. To return to what I was about to say: last night, when I was over at the house—by the way, you can see it from here. You passed it in the car.' He indicated a red roof among poplars some three hundred yards away, the only build-ing in sight that stood separate from the tiny village in the gap below them.

'Certainly I did,' said Trent. 'The manager told me all about it, among other things, as he drove me in from Bishopsbridge.'

'Other people here have heard of you and your performances,' Mr Cupples went on. 'As I was saying, when I was over there last night, Mr Bunner, who is one of Manderson's two secretaries, ex-pressed a hope that the *Record* would send you down to deal with the case, as the police seemed quite at a loss. He mentioned one or two of your past successes, and Mabel—my niece—was interested when I told her afterwards. She is bearing up wonderfully well, Trent; she has remarkable fortitude of character. She said she remembered reading your articles about the Abinger case. She has a great horror of the newspaper side of this sad business, and she had entreated me to do anything I could to keep journalists away from the place— I'm sure you can understand her feeling, Trent; it isn't really any reflection on that profession. But she said you appeared to have great powers as a detective, and she would not stand in the way of anything that might clear up the crime. Then I told her you were a personal friend of mine, and gave you a good character for tact and consideration of others' feelings; and it ended in her saying that, if you should come, she would like you to be helped in every way.'

Trent leaned across the table and shook Mr Cupples by the hand in silence. Mr Cupples, much delighted with the way things were turning out, resumed:

'I spoke to my niece on the telephone only just now, and she is glad you are here. She asks me to say that you may make any enquiries you like, and she puts the house and grounds at your disposal. She had rather not see you herself; she is keeping to her own sitting-room. She has already been interviewed by a detective officer who is there, and she feels unequal to any more. She adds that she does not believe she could say anything that would be of the smallest use. The two secretaries and Martin, the butler (who is a most intelligent man), could tell you all you want to know, she thinks.'

Trent finished his breakfast with a thoughtful brow. He filled a pipe slowly, and seated himself on the rail of the veranda. 'Cupples,' he said quietly, 'is there anything about this business that you know and would rather not tell me?'

Mr Cupples gave a slight start, and turned an astonished gaze on the questioner. 'What do you mean?' he said.

'I mean about the Mandersons. Look here! shall I tell you a thing that strikes me about this affair at the very beginning? Here's a man suddenly and violently killed, and nobody's heart seems to be broken about it, to say the least. The manager of this hotel spoke to me about him as coolly as if he'd never set eyes on him, though I understand they've been neighbours every summer for some years. Then you talk about the thing in the coldest of blood. And Mrs Manderson—well, you won't mind my saying that I have heard of women being more cut up about their husbands being murdered than she seems to be. Is there something in this, Cupples, or is it my fancy? Was there something queer about Manderson? I travelled on the same boat with him once, but never spoke to him. I only know his public character, which was repulsive enough. You see, this may have a bearing on the case; that's the only reason why I ask.'

Mr Cupples took time for thought. He fingered his sparse beard and looked out over the sea. At last he turned to Trent. 'I see no reason,' he said, 'why I shouldn't tell you as between ourselves, my dear fellow. I need not say that this must not be referred to, however distantly. The truth is that nobody really liked Manderson; and I think those who were nearest to him liked him least.'

'Why?' the other interjected.

'Most people found a difficulty in explaining why. In trying to account to myself for my own sensations, I could only put it that one felt in the man a complete absence of the sympathetic faculty. There was nothing outwardly repellent about him. He was not ill-mannered, or vicious, or dull—indeed, he could be remarkably interesting. But I received the impression that there could be no human creature whom he would not sacrifice in the pursuit of his schemes, in his task of imposing himself and his will upon the world. Perhaps that was fanciful, but I think not altogether so. However, the point is that Mabel, I am sorry to say, was very unhappy. I am nearly twice your age, my dear boy, though you always so kindly try to make me feel as if we were contemporaries—I am getting to be an old man, and a great many people have been good enough to confide their matrimonial troubles to me; but I never knew another case like my niece's and her husband's. I have known her since she was a baby, Trent, and I know—you understand, I think, that I do not employ that word lightly—I *know* that she is as amiable and honourable a woman, to say nothing of her other good gifts, as any man could wish. But Manderson, for some time past, had made her miserable.'

'What did he do?' asked Trent, as Mr Cupples paused.

'When I put that question to Mabel, her words were that he seemed to nurse a perpetual grievance. He maintained a distance between them, and he would say nothing. I don't know how it began or what was behind it; and all she would tell me on that point was that he had no cause in the world for his attitude. I think she knew what was in his mind, whatever it was; but she is full of pride. This seems to have gone on for months. At last, a week ago, she wrote to me. I am the only near relative she has. Her mother died when she was a child; and after John Peter died I was something like a father to her until she married—that was five years ago. She asked me to come and help her, and I came at once. That is why I am here now.'

Mr Cupples paused and drank some tea. Trent smoked and stared out at the hot June landscape.

'I would not go to White Gables,' Mr Cupples resumed. 'You know my views, I think, upon the economic constitution of society, and the proper relationship of the capitalist to the employee, and you know, no doubt, what use that person made of his vast industrial power upon several very notorious occasions. I refer especially to the trouble in the Pennsylvania coal-fields, three years ago. I regarded him, apart from all personal dislike, in the light of a criminal and a disgrace to society. I came to this hotel, and I saw my niece here. She told me what I have more briefly told you. She said that the worry and the humiliation of it, and the strain of trying to keep up appearances before the world, were telling upon her, and she asked for my advice. I said I thought she should face him and demand an explanation of his way of treating her. But she would not do that. She had always taken the line of affecting not to notice the change in his demeanour, and nothing, I knew, would persuade her to admit to him that she was injured, once pride had led her into that course. Life is quite full, my dear Trent,' said Mr Cupples with a sigh, 'of these obstinate silences and cultivated misunderstandings.'

'Did she love him?' Trent enquired abruptly. Mr Cupples did not reply at once. 'Had she any love left for him?' Trent amended.

Mr Cupples played with his teaspoon. 'I am bound to say,' he answered slowly, 'that I think not. But you must not misunderstand the woman, Trent. No power on earth would have persuaded her to admit that to any one—even to herself, perhaps—so long as she considered herself bound to him. And I gather that, apart from this mysterious sulking of late, he had always been considerate and generous.'

'You were saying that she refused to have it out with him.'

'She did,' replied Mr Cupples. 'And I knew by experience that it was quite useless to attempt to move a Domecq where the sense of dignity was involved. So I thought it over carefully, and next day I watched my opportunity and met Manderson as he passed by this hotel. I asked him to favour me with a few minutes' conversation, and he stepped inside the gate down there. We had held no communication of any kind since my niece's marriage, but he remembered me, of course. I put the matter to him at once and quite definitely. I told him what Mabel had confided to me. I said that I would neither approve nor condemn her action in bringing me into the business, but that she was suffering, and I considered it my right to ask how he could justify himself in placing her in such a position.'

'And how did he take that?' said Trent, smiling secretly at the landscape. The picture of this mildest of men calling the formidable Manderson to account pleased him.

'Not very well,' Mr Cupples replied sadly. 'In fact, far from well. I can tell you almost exactly what he said—it wasn't much. He said,

"See here, Cupples, you don't want to butt in. My wife can look after herself. I've found that out, along with other things." He was perfectly quiet—you know he was said never to lose control of himself—though there was a light in his eyes that would have frightened a man who was in the wrong, I dare say. But I had been thoroughly roused by his last remark, and the tone of it, which I cannot reproduce. You see,' said Mr Cupples simply, 'I love my niece. She is the only child that there has been in our—in my house. Moreover, my wife brought her up as a girl, and any reflection on Mabel I could not help feeling, in the heat of the moment, as an indirect reflection upon one who is gone.'

'You turned upon him,' suggested Trent in a low tone. 'You asked him to explain his words.'

'That is precisely what I did,' said Mr Cupples. 'For a moment he only stared at me, and I could see a vein on his forehead swelling—an unpleasant sight. Then he said quite quietly, "This thing has gone far enough, I guess," and turned to go.'

'Did he mean your interview?' Trent asked thoughtfully.

'From the words alone you would think so,' Mr Cupples answered. 'But the way in which he uttered them gave me a strange and very apprehensive feeling. I received the impression that the man had formed some sinister resolve. But I regret to say I had lost the power of dispassionate thought. I fell into a great rage'—Mr Cupples's tone was mildly apologetic—'and said a number of foolish things. I reminded him that the law allowed a measure of freedom to wives who received intolerable treatment. I made some utterly irrelevant references to his public record, and expressed the view that such men as he were unfit to live. I said these things, and others as ill-considered, under the eyes, and very possibly within earshot, of half a dozen persons sitting on this veranda. I noticed them, in spite of my agitation, looking at me as I walked up to the hotel again after relieving my mind—for it undoubtedly did relieve it,' sighed Mr Cupples, lying back in his chair.

'And Manderson? Did he say no more?'

'Not a word. He listened to me with his eyes on my face, as quiet as before. When I stopped he smiled very slightly, and at once turned away and strolled through the gate, making for White Gables.'

'And this happened—?'

'On the Sunday morning.'

'Then I suppose you never saw him alive again?'

'No,' said Mr Cupples. 'Or rather yes—once. It was later in the day, on the golf-course. But I did not speak to him. And next morning he was found dead.'

The two regarded each other in silence for a few moments. A party of guests who had been bathing came up the steps and seated themselves, with much chattering, at a table near them. The waiter approached. Mr Cupples rose, and, taking Trent's arm, led him to a long tennis-lawn at the side of the hotel.

'I have a reason for telling you all this,' began Mr Cupples as they paced slowly up and down.

'Trust you for that,' rejoined Trent, carefully filling his pipe again. He lit it, smoked a little, and then said, 'I'll try and guess what your reason is, if you like.'

Mr Cupples's face of solemnity relaxed into a slight smile. He said nothing.

'You thought it possible,' said Trent meditatively—'may I say you thought it practically certain?—that I should find out for myself that there had been something deeper than a mere conjugal tiff between the Mandersons. You thought that my unwholesome imagination would begin at once to play with the idea of Mrs Manderson having something to do with the crime. Rather than that I should lose myself in barren speculations about this, you decided to tell me exactly how matters stood, and incidentally to impress upon me, who know how excellent your judgement is, your opinion of your niece. Is that about right?'

'It is perfectly right. Listen to me, my dear fellow,' said Mr Cupples earnestly, laying his hand on the other's arm. 'I am going to be very frank. I am extremely glad that Manderson is dead. I believe him to have done nothing but harm in the world as an economic factor. I know that he was making a desert of the life of one who was like my own child to me. But I am under an intolerable dread of Mabel being involved in suspicion with regard to the murder. It is horrible to me to think of her delicacy and goodness being in contact, if only for a time, with the brutalities of the law. She is not fitted for it. It would mark her deeply. Many young women of twenty-six in these days could face such an ordeal, I suppose. I have observed a sort of imitative hardness about the products of the higher education of women today which would carry them through anything, perhaps.

I am not prepared to say it is a bad thing in the conditions of feminine life prevailing at present. Mabel, however, is not like that. She is as unlike that as she is unlike the simpering misses that used to surround me as a child. She has plenty of brains; she is full of character; her mind and her tastes are cultivated; but it is all mixed up'— Mr Cupples waved his hands in a vague gesture—'with ideals of refinement and reservation and womanly mystery. I fear she is not a child of the age. You never knew my wife, Trent. Mabel is my wife's child.'

The younger man bowed his head. They paced the length of the lawn before he asked gently, 'Why did she marry him?'

'I don't know,' said Mr Cupples briefly.

'Admired him, I suppose,' suggested Trent.

Mr Cupples shrugged his shoulders. 'I have been told that a woman will usually be more or less attracted by the most successful man in her circle. Of course we cannot realize how a wilful, dominating personality like his would influence a girl whose affections were not bestowed elsewhere; especially if he laid himself out to win her. It is probably an overwhelming thing to be courted by a man whose name is known all over the world. She had heard of him, of course, as a financial great power, and she had no idea—she had lived mostly among people of artistic or literary propensities—how much soulless inhumanity that might involve. For all I know, she has no adequate idea of it to this day. When I first heard of the affair the mischief was done, and I knew better than to interpose my unsought opinions. She was of age, and there was absolutely nothing against him from the conventional point of view. Then I dare say his immense wealth would cast a spell over almost any woman. Mabel had some hundreds a year of her own; just enough, perhaps, to let her realize what millions really meant. But all this is conjecture. She certainly had not wanted to marry some scores of young fellows who to my knowledge had asked her; and though I don't believe, and never did believe, that she really loved this man of forty-five, she certainly did want to marry him. But if you ask me why, I can only say I don't know.'

Trent nodded, and after a few more paces looked at his watch. 'You've interested me so much,' he said, 'that I had quite forgotten my main business. I mustn't waste my morning. I am going down the road to White Gables at once, and I dare say I shall be poking about there until midday. If you can meet me then, Cupples, I should like to talk over anything I find out with you, unless something detains me.'

'I am going for a walk this morning,' Mr Cupples replied. 'I meant to have luncheon at a little inn near the golf-course, The Three Tuns. You had better join me there. It's further along the road, about a quarter of a mile beyond White Gables. You can just see the roof between those two trees. The food they give one there is very plain, but good.'

'So long as they have a cask of beer,' said Trent, 'they are all right. We will have bread and cheese, and oh, may Heaven our simple lives prevent from luxury's contagion, weak and vile! Till then, goodbye.' He strode off to recover his hat from the veranda, waved it to Mr Cupples, and was gone.

The old gentleman, seating himself in a deck-chair on the lawn, clasped his hands behind his head and gazed up into the speckless blue sky. 'He is a dear fellow,' he murmured. 'The best of fellows. And a terribly acute fellow. Dear me! How curious it all is!'

CHAPTER IV

Handcuffs in the Air

A PAINTER and the son of a painter, Philip Trent had while yet in his twenties achieved some reputation within the world of English art. Moreover, his pictures sold. An original, forcible talent and a habit of leisurely but continuous working, broken by fits of strong creative enthusiasm, were at the bottom of it. His father's name had helped; a patrimony large enough to relieve him of the perilous imputation of being a struggling man had certainly not hindered. But his best aid to success had been an unconscious power of getting himself liked. Good spirits and a lively, humorous fancy will always be popular. Trent joined to these a genuine interest in others that gained him something deeper than popularity. His judgement of persons was penetrating, but its process was internal; no one felt on good behaviour with a man who seemed always to be enjoying himself. Whether he was in a mood for floods of nonsense or applying himself vigorously to a task, his face seldom lost its expression of contained vivacity. Apart from a sound knowledge of his art and its history, his culture was large and loose, dominated by a love of poetry. At thirty-two he had not yet passed the age of laughter and adventure.

His rise to a celebrity a hundred times greater than his proper work had won for him came of a momentary impulse. One day he had taken up a newspaper to find it chiefly concerned with a crime of a sort curiously rare in our country—a murder done in a railway train. The circumstances were puzzling; two persons were under arrest upon suspicion. Trent, to whom an interest in such affairs was a new sensation, heard the thing discussed among his friends, and set himself in a purposeless mood to read up the accounts given in several journals. He became intrigued; his imagination began to work, in a manner strange to him, upon facts; an excitement took hold of him such as he had only known before in his bursts of art-inspiration or of personal adventure. At the end of the day he wrote and dispatched a long letter to the editor of the *Record*, which he chose only because it had contained the fullest and most intelligent version of the facts.

In this letter he did very much what Poe had done in the case of the murder of Mary Rogers. With nothing but the newspapers to guide him, he drew attention to the significance of certain apparently negligible facts, and ranged the evidence in such a manner as to throw grave suspicion upon a man who had presented himself as a witness. Sir James Molloy had printed this letter in leaded type. The same evening he was able to announce in the *Sun* the arrest and full confession of the incriminated man.

Sir James, who knew all the worlds of London, had lost no time in making Trent's acquaintance. The two men got on well, for Trent possessed some secret of native tact which had the effect of almost abolishing differences of age between himself and others. The great rotary presses in the basement of the *Record* building had filled him with a new enthusiasm. He had painted there, and Sir James had bought at sight, what he called a machinery-scape in the manner of Heinrich Kley.

Then a few months later came the affair known as the Ilkley mystery. Sir James had invited Trent to an emollient dinner, and thereafter offered him what seemed to the young man a fantastically large sum for his temporary services as special representative of the *Record* at Ilkley.

'You could do it,' the editor had urged. 'You can write good stuff, and you know how to talk to people, and I can teach you all the technicalities of a reporter's job in half an hour. And you have a head for a mystery; you have imagination and cool judgement along with it. Think how it would feel if you pulled it off!'

Trent had admitted that it would be rather a lark. He had smoked, frowned, and at last convinced himself that the only thing that held him back was fear of an unfamiliar task. To react against fear had become a fixed moral habit with him, and he had accepted Sir James's offer.

He had pulled it off. For the second time he had given the authorities a start and a beating, and his name was on all tongues. He withdrew and painted pictures. He felt no leaning towards journalism, and Sir James, who knew a good deal about art, honourably refrained—as other editors did not—from tempting him with a good salary. But in the course of a few years he had applied to him perhaps thirty times for his services in the unravelling of similar problems at home and abroad. Sometimes Trent, busy with work that held him, had refused; sometimes he had been forestalled in the discovery of the truth. But the result of his irregular connection with the *Record* had been to make his name one of the best known in England. It was characteristic of him that his name was almost the only detail of his personality known to the public. He had imposed absolute silence about himself upon the Molloy papers; and the others were not going to advertise one of Sir James's men.

The Manderson case, he told himself as he walked rapidly up the sloping road to White Gables, might turn out to be terribly simple. Cupples was a wise old boy, but it was probably impossible for him to have an impartial opinion about his niece. But it was true that the manager of the hotel, who had spoken of her beauty in terms that aroused his attention, had spoken even more emphatically of her goodness. Not an artist in words, the manager had yet conveyed a very definite idea to Trent's mind. 'There isn't a child about here that don't brighten up at the sound of her voice,' he had said, 'nor yet a grown-up, for the matter of that. Everybody used to look forward to her coming over in the summer. I don't mean that she's one of those women that are all kind heart and nothing else. There's backbone with it, if you know what I mean—pluck—any amount of go. There's nobody in Marlstone that isn't sorry for the lady in her trouble—not but what some of us may think she's lucky at the last of it.' Trent wanted very much to meet Mrs Manderson.

He could see now, beyond a spacious lawn and shrubbery, the front of the two-storied house of dull-red brick, with the pair of great gables from which it had its name. He had had but a glimpse of it from the car that morning. A modern house, he saw; perhaps ten years old. The place was beautifully kept, with that air of opulent peace that clothes even the smallest houses of the well-to-do in an English countryside. Before it, beyond the road, the rich meadow-land ran down to the edge of the cliffs; behind it a woody landscape stretched away across a broad vale to the moors. That such a place could be the scene of a crime of violence seemed fantastic; it lay so quiet and well ordered, so eloquent of disciplined service and gentle living. Yet there beyond the house, and near the hedge that rose between the garden and the hot, white road, stood the gardener's toolshed, by which the body had been found, lying tumbled against the wooden wall.

Trent walked past the gate of the drive and along the road until he was opposite this shed. Some forty yards further along the road turned sharply away from the house, to run between thick plantations; and just before the turn the grounds of the house ended, with a small white gate at the angle of the boundary hedge. He approached the gate, which was plainly for the use of gardeners and the service of the establishment. It swung easily on its hinges, and he passed slowly up a path that led towards the back of the house, between the outer hedge and a tall wall of rhododendrons. Through a gap in this wall a track led him to the little neatly built erection of wood, which stood among trees that faced a corner of the front. The body had lain on the side away from the house; a servant, he thought, looking out of the nearer windows in the earlier hours of the day before, might have glanced unseeing at the hut, as she wondered what it could be like to be as rich as the master.

He examined the place carefully and ransacked the hut within, but he could note no more than the trodden appearance of the uncut grass where the body had lain. Crouching low, with keen eyes and feeling fingers, he searched the ground minutely over a wide area; but the search was fruitless.

It was interrupted by the sound—the first he had heard from the house—of the closing of the front door. Trent unbent his long legs and stepped to the edge of the drive. A man was walking quickly away from the house in the direction of the great gate.

At the noise of a footstep on the gravel, the man wheeled with nervous swiftness and looked earnestly at Trent. The sudden sight of his face was almost terrible, so white and worn it was. Yet it was a young man's face. There was not a wrinkle about the haggard blue eyes, for all their tale of strain and desperate fatigue. As the two approached each other, Trent noted with admiration the man's breadth of shoulder and lithe, strong figure. In his carriage, inelastic as weariness had made it; in his handsome, regular features; in his short, smooth, yellow hair; and in his voice as he addressed Trent, the influence of a special sort of training was confessed. 'Oxford was your playground, I think, my young friend,' said Trent to himself.

'If you are Mr Trent,' said the young man pleasantly, 'you are expected. Mr Cupples telephoned from the hotel. My name is Marlowe.'

'You were secretary to Mr Manderson, I believe,' said Trent. He was much inclined to like young Mr Marlowe. Though he seemed so near a physical breakdown, he gave out none the less that air of clean living and inward health that is the peculiar glory of his social type at his years. But there was something in the tired eyes that was a challenge to Trent's penetration; an habitual expression, as he took it to be, of meditating and weighing things not present to their sight. It was a look too intelligent, too steady and purposeful, to be called dreamy. Trent thought he had seen such a look before somewhere. He went on to say: 'It is a terrible business for all of you. I fear it has upset you completely, Mr Marlowe.'

'A little limp, that's all,' replied the young man wearily. 'I was driving the car all Sunday night and most of yesterday, and I didn't sleep last night after hearing the news—who would? But I have an appointment now, Mr Trent, down at the doctor's—arranging about the inquest. I expect it'll be tomorrow. If you will go up to the house and ask for Mr Bunner, you'll find him expecting you; he will tell you all about things and show you round. He's the other secretary; an American, and the best of fellows; he'll look after you. There's a detective here, by the way—Inspector Murch, from Scotland Yard. He came yesterday.'

'Murch!' Trent exclaimed. 'But he and I are old friends. How under the sun did he get here so soon?'

'I have no idea,' Mr Marlowe answered. 'But he was here last evening, before I got back from Southampton, interviewing everybody, and he's been about here since eight this morning. He's in the library now—that's where the open French window is that you see at the end of the house there. Perhaps you would like to step down there and talk about things.'

'I think I will,' said Trent. Marlowe nodded and went on his way. The thick turf of the lawn round which the drive took its circular sweep made Trent's footsteps as noiseless as a cat's. In a few moments he was looking in through the open leaves of the window at the southward end of the house, considering with a smile a very broad back and a bent head covered with short grizzled hair. The man within was stooping over a number of papers laid out on the table.

' 'Twas ever thus,' said Trent in a melancholy tone, at the first sound of which the man within turned round with startling swiftness. 'From childhood's hour I've seen my fondest hopes decay. I did think I was ahead of Scotland Yard this time, and now here is the largest officer in the entire Metropolitan force already occupying the position.'

The detective smiled grimly and came to the window. 'I was expecting you, Mr Trent,' he said. 'This is the sort of case that you like.'

'Since my tastes were being considered,' Trent replied, stepping into the room, 'I wish they had followed up the idea by keeping my hated rival out of the business. You have got a long start, too—I know all about it.' His eyes began to wander round the room. 'How did you manage it? You are a quick mover, I know; the dun deer's hide on fleeter foot was never tied; but I don't see how you got here in time to be at work yesterday evening. Has Scotland Yard secretly started an aviation corps? Or is it in league with the infernal powers? In either case the Home Secretary should be called upon to make a statement.'

'It's simpler than that,' said Mr Murch with professional stolidity. 'I happened to be on leave with the missis at Halvey, which is only twelve miles or so along the coast. As soon as our people there heard of the murder they told me. I wired to the Chief, and was put in charge of the case at once. I bicycled over yesterday evening, and have been at it since then.'

'Arising out of that reply,' said Trent inattentively, 'how is Mrs Inspector Murch?'

'Never better, thank you,' answered the inspector, 'and frequently speaks of you and the games you used to have with our kids. But you'll excuse me saying, Mr Trent, that you needn't trouble to talk your nonsense to me while you're using your eyes. I know your ways by now. I understand you've fallen on your feet as usual, and have the lady's permission to go over the place and make enquiries.'

'Such is the fact,' said Trent. 'I am going to cut you out again, inspector. I owe you one for beating me over the Abinger case, you old fox. But if you really mean that you're not inclined for the social amenities just now, let us leave compliments and talk business.' He stepped to the table, glanced through the papers arranged there in order, and then turned to the open roll-top desk. He looked into the drawers swiftly. 'I see this has been cleared out. Well now, inspector, I suppose we play the game as before.'

Trent had found himself on a number of occasions in the past thrown into the company of Inspector Murch, who stood high in the councils of the Criminal Investigation Department. He was a quiet, tactful, and very shrewd officer, a man of great courage, with a vivid history in connection with the more dangerous class of criminals. His humanity was as broad as his frame, which was large even for a policeman. Trent and he, through some obscure working of sympathy, had appreciated one another from the beginning, and had formed one of those curious friendships with which it was the younger man's delight to adorn his experience. The inspector would talk more freely to him than to any one, under the rose, and they would discuss details and possibilities of every case, to their mutual enlightenment. There were necessarily rules and limits. It was understood between them that Trent made no journalistic use of any point that could only have come to him from an official source. Each of them, moreover, for the honour and prestige of the institution he represented, openly reserved the right to withhold from the other any discovery or inspiration that might come to him which he considered vital to the solution of the difficulty. Trent had insisted on carefully formulating these principles of what he called detective sportsmanship. Mr Murch, who loved a contest, and who only stood to gain by his association with the keen intelligence of the other, entered very heartily into 'the game'. In these strivings for the credit of the press and of the police, victory sometimes attended the experience and method of the officer, sometimes the quicker brain and livelier imagination of Trent, his gift of instinctively recognizing the significant through all disguises.

The inspector then replied to Trent's last words with cordial agreement. Leaning on either side of the French window, with the deep peace and hazy splendour of the summer landscape before them, they reviewed the case.

Trent had taken out a thin notebook, and as they talked he began to make, with light, secure touches, a rough sketch plan of the room. It was a thing he did habitually on such occasions, and often quite idly, but now and then the habit had served him to good purpose.

This was a large, light apartment at the corner of the house, with generous window-space in two walls. A broad table stood in the middle. As one entered by the window the roll-top desk stood just to the left of it against the wall. The inner door was in the wall to the left, at the farther end of the room; and was faced by a broad window divided into openings of the casement type. A beautifully carved old corner-cupboard rose high against the wall beyond the door, and another cupboard filled a recess beside the fireplace. Some coloured prints of Harunobu, with which Trent promised himself a better acquaintance, hung on what little wall-space was unoccupied by books. These had a very uninspiring appearance of having been bought by the yard and never taken from their shelves. Bound with a sober luxury, the great English novelists, essayists, historians, and poets stood ranged like an army struck dead in its ranks. There were a few chairs made, like the cupboard and table, of old carved oak; a modern armchair and a swivel office-chair before the desk. The room looked costly but very bare. Almost the only portable objects were a great porcelain bowl of a wonderful blue on the table, a clock and some cigar boxes on the mantelshelf, and a movable telephone standard on the top of the desk.

'Seen the body?' enquired the inspector.

Trent nodded. 'And the place where it lay,' he said.

'First impressions of this case rather puzzle me,' said the inspector. 'From what I heard at Halvey I guessed it might be common robbery and murder by some tramp, though such a thing is very far from common in these parts. But as soon as I began my enquiries I came on some curious points, which by this time I dare say you've noted for yourself. The man is shot in his own grounds, quite near the house, to begin with. Yet there's not the slightest trace of any attempt at burglary. And the body wasn't robbed. In fact, it would be as plain a case of suicide as you could wish to see, if it wasn't for certain facts. Here's another thing: for a month or so past, they tell me, Manderson had been in a queer state of mind. I expect you know already that he and his wife had some trouble between them. The servants had noticed a change in his manner to her for a long time, and for the past week he had scarcely spoken to her. They say he was a changed man, moody and silent—whether on account of that or something else. The lady's maid says he looked as if something was going to arrive. It's always easy to remember that people looked like that, after something has happened to them. Still, that's what they say. There you are again, then: suicide! Now, why wasn't it suicide, Mr Trent?'

'The facts so far as I know them are really all against it,' Trent replied, sitting on the threshold of the window and clasping his knees. 'First, of course, no weapon is to be found. I've searched, and you've searched, and there's no trace of any firearm anywhere within a stone's throw of where the body lay. Second, the marks on the wrists, fresh scratches and bruises, which we can only assume to have been done in a struggle with somebody. Third, who ever heard of anybody shooting himself in the eye? Then I heard from the manager of the hotel here another fact, which strikes me as the most curious detail in this affair. Manderson had dressed himself fully before going out there, but he forgot his false teeth. Now how could a suicide who dressed himself to make a decent appearance as a corpse forget his teeth?'

'That last argument hadn't struck me,' admitted Mr Murch. 'There's something in it. But on the strength of the other points, which had occurred to me, I am not considering suicide. I have been looking about for ideas in this house, this morning. I expect you were thinking of doing the same.'

'That is so. It is a case for ideas, it seems to me. Come, Murch, let us make an effort; let us bend our spirits to a temper of general suspicion. Let us suspect everybody in the house, to begin with. Listen: I will tell you whom I suspect. I suspect Mrs Manderson, of course. I also suspect both the secretaries—I hear there are two, and I hardly know which of them I regard as more thoroughly open to suspicion. I suspect the butler and the lady's maid. I suspect the other domestics, and especially do I suspect the boot-boy. By the way, what domestics are there? I have more than enough suspicion to go round, whatever the size of the establishment; but as a matter of curiosity I should like to know.'

'All very well to laugh,' replied the inspector, 'but at the first stage of affairs it's the only safe principle, and you know that as well as I do, Mr Trent. However, I've seen enough of the people here, last night and today, to put a few of them out of my mind for the present at least. You will form your own conclusions. As for the establishment, there's the butler and lady's maid, cook, and three other maids, one a young girl. One chauffeur, who's away with a broken wrist. No boy.'

'What about the gardener? You say nothing about that shadowy and sinister figure, the gardener. You are keeping him in the background, Murch. Play the game. Out with him—or I report you to the Rules Committee.'

'The garden is attended to by a man in the village, who comes twice a week. I've talked to him. He was here last on Friday.'

'Then I suspect him all the more,' said Trent. 'And now as to the house itself. What I propose to do, to begin with, is to sniff about a little in this room, where I am told Manderson spent a great deal of his time, and in his bedroom; especially the bedroom. But since we're in this room, let's start here. You seem to be at the same stage of the enquiry. Perhaps you've done the bedrooms already?'

The inspector nodded. 'I've been over Manderson's and his wife's. Nothing to be got there, I think. His room is very simple and bare, no signs of any sort—that *I* could see. Seems to have insisted on the simple life, does Manderson. Never employed a valet. The room's almost like a cell, except for the clothes and shoes. You'll find it all exactly as I found it; and they tell me that's exactly as Manderson left it, at we don't know what o'clock yesterday morning. Opens into Mrs Manderson's bedroom—not much of the cell about that, I can tell you. I should say the lady was as fond of pretty things as most. But she cleared out of it on the morning of the discovery—told the maid she could never sleep in a room opening into her murdered husband's room. Very natural feeling in a woman, Mr Trent. She's camping out, so to say, in one of the spare bedrooms now.'

'Come, my friend,' Trent was saying to himself, as he made a few notes in his little book. 'Have you got your eye on Mrs Manderson? Or haven't you? I know that colourless tone of the inspectorial voice. I wish I had seen her. Either you've got something against her and you don't want me to get hold of it; or else you've made up your mind she's innocent, but have no objection to my wasting my time over her. Well, it's all in the game; which begins to look extremely interesting as we go on.' To Mr Murch he said aloud: 'Well, I'll draw the bedroom later on. What about this?'

'They call it the library,' said the inspector. 'Manderson used to do his writing and that in here; passed most of the time he spent indoors here . Since he and his wife ceased to hit it off together, he had taken to spending his evenings alone, and when at this house he always spent 'em in here. He was last seen alive, as far as the servants are concerned, in this room.'

Trent rose and glanced again through the papers set out on the table. 'Business letters and documents, mostly,' said Mr Murch. 'Reports, prospectuses, and that. A few letters on private matters, nothing in them that I can see. The American secretary—Bunner his name is, and a queerer card I never saw turned—he's been through this desk with me this morning. He had got it into his head that Manderson had been receiving threatening letters, and that the murder was the outcome of that. But there's no trace of any such thing;

and we looked at every blessed paper. The only unusual things we found were some packets of banknotes to a considerable amount, and a couple of little bags of unset diamonds. I asked Mr Bunner to put them in a safer place. It appears that Manderson had begun buying diamonds lately as a speculation—it was a new game to him, the secretary said, and it seemed to amuse him.'

'What about these secretaries?' Trent enquired. 'I met one called Marlowe just now outside; a nice-looking chap with singular eyes, unquestionably English. The other, it seems, is an American. What did Manderson want with an English secretary?'

'Mr Marlowe explained to me how that was. The American was his right-hand business man, one of his office staff, who never left him. Mr Marlowe had nothing to do with Manderson's business as a financier, knew nothing of it. His job was to look after Manderson's horses and motors and yacht and sporting arrangements and that— make himself generally useful, as you might say. He had the spending of a lot of money, I should think. The other was confined entirely to the office affairs, and I dare say he had his hands full. As for his being English, it was just a fad of Manderson's to have an English secretary. He'd had several before Mr Marlowe.'

'He showed his taste,' observed Trent. 'It might be more than interesting, don't you think, to be minister to the pleasures of a modern plutocrat with a large P. Only they say that Manderson's were exclusively of an innocent kind. Certainly Marlowe gives me the impression that he would be weak in the part of Petronius. But to return to the matter in hand.' He looked at his notes. 'You said just now that he was last seen alive here, "so far as the servants were concerned". That meant—?'

'He had a conversation with his wife on going to bed. But for that, the manservant, Martin by name, last saw him in this room. I had his story last night, and very glad he was to tell it. An affair like this is meat and drink to the servants of the house.'

Trent considered for some moments, gazing through the open window over the sun-flooded slopes. 'Would it bore you to hear what he has to say again?' he asked at length. For reply, Mr Murch rang the bell. A spare, clean-shaven, middle-aged man, having the servant's manner in its most distinguished form, answered it.

'This is Mr Trent, who is authorized by Mrs Manderson to go over the house and make enquiries,' explained the detective. 'He would like to hear your story.' Martin bowed distantly. He recognized Trent for a gentleman. Time would show whether he was what Martin called a gentleman in every sense of the word.

'I observed you approaching the house, sir,' said Martin with impassive courtesy. He spoke with a slow and measured utterance. 'My instructions are to assist you in every possible way. Should you wish me to recall the circumstances of Sunday night?'

'Please,' said Trent with ponderous gravity. Martin's style was making clamorous appeal to his sense of comedy. He banished with an effort all vivacity of expression from his face.

'I last saw Mr Manderson—'

'No, not that yet,' Trent checked him quietly. 'Tell me all you saw of him that evening—after dinner, say. Try to recollect every little detail.'

'After dinner, sir?—yes. I remember that after dinner Mr Manderson and Mr Marlowe walked up and down the path through the orchard, talking. If you ask me for details, it struck me they were talking about something important, because I heard Mr Manderson say something when they came in through the back entrance. He said, as near as I can remember, "If Harris is there, every minute is of importance. You want to start right away. And not a word to a soul." Mr Marlowe answered, "Very well. I will just change out of these clothes and then I am ready"—or words to that effect. I heard this plainly as they passed the window of my pantry. Then Mr Marlowe went up to his bedroom, and Mr Manderson entered the library and rang for me. He handed me some letters for the postman in the morning and directed me to sit up, as Mr Marlowe had persuaded him to go for a drive in the car by moonlight.'

'That was curious,' remarked Trent.

'I thought so, sir. But I recollected what I had heard about "not a word to a soul", and I concluded that this about a moonlight drive was intended to mislead.'

'What time was this?'

'It would be about ten, sir, I should say. After speaking to me, Mr Manderson waited until Mr Marlowe had come down and brought round the car. He then went into the drawing-room, where Mrs Manderson was.'

'Did that strike you as curious?'

Martin looked down his nose. 'If you ask me the question, sir,' he said with reserve, 'I had not known him enter that room since we came here this year. He preferred to sit in the library in the evenings.

That evening he only remained with Mrs Manderson for a few minutes. Then he and Mr Marlowe started immediately.'

'You saw them start?'

'Yes, sir. They took the direction of Bishopsbridge.'

'And you saw Mr Manderson again later?'

'After an hour or thereabouts, sir, in the library. That would have been about a quarter past eleven, I should say; I had noticed eleven striking from the church. I may say I am peculiarly quick of hearing, sir.'

'Mr Manderson had rung the bell for you, I suppose. Yes? And what passed when you answered it?'

'Mr Manderson had put out the decanter of whisky and a syphon and glass, sir, from the cupboard where he kept them—'

Trent held up his hand. 'While we are on that point, Martin, I want to ask you plainly, did Mr Manderson drink very much? You understand this is not impertinent curiosity on my part. I want you to tell me, because it may possibly help in the clearing up of this case.'

'Perfectly, sir,' replied Martin gravely. 'I have no hesitation in telling you what I have already told the inspector. Mr Manderson was, considering his position in life, a remarkably abstemious man. In my four years of service with him I never knew anything of an alcoholic nature pass his lips, except a glass or two of wine at dinner, very rarely a little at luncheon, and from time to time a whisky and soda before going to bed. He never seemed to form a habit of it. Often I used to find his glass in the morning with only a little soda water in it; sometimes he would have been having whisky with it, but never much. He never was particular about his drinks; ordinary soda was what he preferred, though I had ventured to suggest some of the natural minerals, having personally acquired a taste for them in my previous service. He used to keep them in the cupboard here, because he had a great dislike of being waited on more than was necessary. It was an understood thing that I never came near him after dinner unless sent for. And when he sent for anything, he liked it brought quick, and to be left alone again at once. He hated to be asked if he required anything more. Amazingly simple in his tastes, sir, Mr Manderson was.'

'Very well; and he rang for you that night about a quarter past eleven. Now can you remember exactly what he said?'

'I think I can tell you with some approach to accuracy, sir. It was not much. First he asked me if Mr Bunner had gone to bed, and I replied that he had been gone up some time. He then said that he wanted some one to sit up until 12.30, in case an important message should come by telephone, and that Mr Marlowe having gone to Southampton for him in the motor, he wished me to do this, and that I was to take down the message if it came, and not disturb him. He also ordered a fresh syphon of soda water. I believe that was all, sir.'

'You noticed nothing unusual about him, I suppose?'

'No, sir, nothing unusual. When I answered the ring, he was seated at the desk listening at the telephone, waiting for a number, as I supposed. He gave his orders and went on listening at the same time. When I returned with the syphon he was engaged in conversation over the wire.'

'Do you remember anything of what he was saying?'

'Very little, sir; it was something about somebody being at some hotel—of no interest to me. I was only in the room just time enough to place the syphon on the table and withdraw. As I closed the door he was saying, "You're sure he isn't in the hotel?" or words to that effect.'

'And that was the last you saw and heard of him alive?'

'No, sir. A little later, at half-past eleven, when I had settled down in my pantry with the door ajar, and a book to pass the time, I heard Mr Manderson go upstairs to bed. I immediately went to close the library window, and slipped the lock of the front door. I did not hear anything more.'

Trent considered. 'I suppose you didn't doze at all,' he said tentatively, 'while you were sitting up waiting for the telephone message?'

'Oh no, sir. I am always very wakeful about that time. I'm a bad sleeper, especially in the neighbourhood of the sea, and I generally read in bed until somewhere about midnight.'

'And did any message come?'

'No, sir.'

'No. And I suppose you sleep with your window open, these warm nights?'

'It is never closed at night, sir.'

Trent added a last note; then he looked thoughtfully through those he had taken. He rose and paced up and down the room for some moments with a downcast eye. At length he paused opposite Martin.

'It all seems perfectly ordinary and simple,' he said. 'I just want to get a few details clear. You went to shut the windows in the library before going to bed. Which windows?'

'The French window, sir. It had been open all day. The windows opposite the door were seldom opened.'

'And what about the curtains? I am wondering whether any one outside the house could have seen into the room.'

'Easily, sir, I should say, if he had got into the grounds on that side. The curtains were never drawn in the hot weather. Mr Manderson would often sit right in the doorway at nights, smoking and looking out into the darkness. But nobody could have seen him who had any business to be there.'

'I see. And now tell me this. Your hearing is very acute, you say, and you heard Mr Manderson enter the house when he came in after dinner from the garden. Did you hear him re-enter it after returning from the motor drive?'

Martin paused. 'Now you mention it, sir, I remember that I did not. His ringing the bell in this room was the first I knew of his being back. I should have heard him come in, if he had come in by the front. I should have heard the door go. But he must have come in by the window.' The man reflected for a moment, then added, 'As a general rule, Mr Manderson would come in by the front, hang up his hat and coat in the hall, and pass down the hall into the study. It seems likely to me that he was in a great hurry to use the telephone, and so went straight across the lawn to the window—he was like that, sir, when there was anything important to be done. He had his hat on, now I remember, and had thrown his greatcoat over the end of the table. He gave his order very sharp, too, as he always did when busy. A very precipitate man indeed was Mr Manderson; a hustler, as they say.'

'Ah! he appeared to be busy. But didn't you say just now that you noticed nothing unusual about him?'

A melancholy smile flitted momentarily over Martin's face. 'That observation shows that you did not know Mr Manderson, sir, if you will pardon my saying so. His being like that was nothing unusual; quite the contrary. It took me long enough to get used to it. Either he would be sitting quite still and smoking a cigar, thinking or reading, or else he would be writing, dictating, and sending off wires all at the same time, till it almost made one dizzy to see it, sometimes for an hour or more at a stretch. As for being in a hurry over a telephone message, I may say it wasn't in him to be anything else.'

Trent turned to the inspector, who met his eye with a look of answering intelligence. Not sorry to show his understanding of the line of inquiry opened by Trent, Mr Murch for the first time put a question.

'Then you left him telephoning by the open window, with the lights on, and the drinks on the table; is that it?'

'That is so, Mr Murch.' The delicacy of the change in Martin's manner when called upon to answer the detective momentarily distracted Trent's appreciative mind. But the big man's next question brought it back to the problem at once.

'About those drinks. You say Mr Manderson often took no whisky before going to bed. Did he have any that night?'

'I could not say. The room was put to rights in the morning by one of the maids, and the glass washed, I presume, as usual. I know that the decanter was nearly full that evening. I had refilled it a few days before, and I glanced at it when I brought the fresh syphon, just out of habit, to make sure there was a decent-looking amount.'

The inspector went to the tall corner-cupboard and opened it. He took out a decanter of cut glass and set it on the table before Martin. 'Was it fuller than that?' he asked quietly. 'That's how I found it this morning.' The decanter was more than half empty.

For the first time Martin's self-possession wavered. He took up the decanter quickly, tilted it before his eyes, and then stared amazedly at the others. He said slowly: 'There's not much short of half a bottle gone out of this since I last set eyes on it—and that was that Sunday night.'

'Nobody in the house, I suppose?' suggested Trent discreetly.

'Out of the question!' replied Martin briefly; then he added, 'I beg pardon, sir, but this is a most extraordinary thing to me. Such a thing never happened in all my experience of Mr Manderson. As for the women-servants, they never touch anything, I can answer for it; and as for me, when I want a drink I can help myself without going to the decanters.' He took up the decanter again and aimlessly renewed his observation of the contents, while the inspector eyed him with a look of serene satisfaction, as a master contemplates his handiwork.

Trent turned to a fresh page of his notebook, and tapped it thoughtfully with his pencil. Then he looked up and said, 'I suppose Mr Manderson had dressed for dinner that night?'

'Certainly, sir. He had on a suit with a dress-jacket, what he used to refer to as a Tuxedo, which he usually wore when dining at home.'

'And he was dressed like that when you saw him last ?'

'All but the jacket, sir. When he spent the evening in the library, as usually happened, he would change it for an old shooting-jacket after dinner, a light-coloured tweed, a little too loud in pattern for English tastes, perhaps. He had it on when I saw him last. It used to hang in this cupboard here'—Martin opened the door of it as he spoke— 'along with Mr Manderson's fishing-rods and such things, so that he could slip it on after dinner without going upstairs.'

'Leaving the dinner-jacket in the cupboard?'

'Yes, sir. The housemaid used to take it upstairs in the morning.'

'In the morning,' Trent repeated slowly. 'And now that we are speaking of the morning, will you tell me exactly what you know about that? I understand that Mr Manderson was not missed until the body was found about ten o'clock.'

'That is so, sir. Mr Manderson would never be called, or have anything brought to him in the morning. He occupied a separate bedroom. Usually he would get up about eight and go round to the bathroom, and he would come down some time before nine. But often he would sleep till nine or ten o'clock. Mrs Manderson was always called at seven. The maid would take in tea to her. Yesterday morning Mrs Manderson took breakfast about eight in her sitting-room as usual, and every one supposed that Mr Manderson was still in bed and asleep, when Evans came rushing up to the house with the shocking intelligence.'

'I see,' said Trent. 'And now another thing. You say you slipped the lock of the front door before going to bed. Was that all the locking-up you did?'

'To the front door, sir, yes; I slipped the lock. No more is considered necessary in these parts. But I had locked both the doors at the back, and seen to the fastenings of all the windows on the ground floor. In the morning everything was as I had left it.'

'As you had left it. Now here is another point—the last, I think. Were the clothes in which the body was found the clothes that Mr Manderson would naturally have worn that day?'

Martin rubbed his chin. 'You remind me how surprised I was when I first set eyes on the body, sir. At first I couldn't make out what was unusual about the clothes, and then I saw what it was. The collar was a shape of collar Mr Manderson never wore except with evening dress. Then I found that he had put on all the same things that he had worn the night before—large fronted shirt and all—except just the coat and waistcoat and trousers, and the brown shoes, and blue tie. As for the suit, it was one of half a dozen he might have worn. But for him to have simply put on all the rest just because they were there, instead of getting out the kind of shirt and things he always wore by day; well, sir, it was unprecedented. It shows, like some other things, what a hurry he must have been in when getting up.'

'Of course,' said Trent. 'Well, I think that's all I wanted to know. You have put everything with admirable clearness, Martin. If we want to ask any more questions later on, I suppose you will be somewhere about.'

'I shall be at your disposal, sir.' Martin bowed, and went out quietly.

Trent flung himself into the armchair and exhaled a long breath. 'Martin is a great creature,' he said. 'He is far, far better than a play. There is none like him, none, nor will be when our summers have deceased. Straight, too; not an atom of harm in dear old Martin. Do you know, Murch, you are wrong in suspecting that man.'

'I never said a word about suspecting him.' The inspector was taken aback. '*You* know, Mr Trent, he would never have told his story like that if he thought I suspected him.'

'I dare say he doesn't think so. He is a wonderful creature, a great artist; but, in spite of that, he is not at all a sensitive type. It has never occurred to his mind that you, Murch, could suspect him, Martin, the complete, the accomplished. But I know it. You must understand, inspector, that I have made a special study of the psychology of officers of the law. It is a grossly neglected branch of knowledge. They are far more interesting than criminals, and not nearly so easy. All the time I was questioning him I saw handcuffs in your eye. Your lips were mutely framing the syllables of those tremendous words: "It is my duty to tell you that anything you now say will be taken down and used in evidence against you." Your manner would have deceived most men, but it could not deceive me.'

Mr Murch laughed heartily. Trent's nonsense never made any sort of impression on his mind, but he took it as a mark of esteem, which indeed it was; so it never failed to please him. 'Well, Mr Trent,' he said, 'you're perfectly right. There's no point in denying it, I have got my eye on him. Not that there's anything definite; but you know as well as I do how often servants are mixed up in affairs of this kind, and this man is such a very quiet customer. You remember the case of Lord William Russell's valet, who went in as usual, in the morning, to draw up the blinds in his master's bedroom, as quiet and starchy as you please, a few hours after he had murdered him in his bed. I've talked to all the women of the house, and I don't believe there's a morsel of harm in one of them. But Martin's not so easy set aside. I don't like his manner; I believe he's hiding something. If so, I shall find it out.'

'Cease!' said Trent. 'Drain not to its dregs the urn of bitter prophecy. Let us get back to facts. Have you, as a matter of evidence, anything at all to bring against Martin's story as he has told it to us?'

'Nothing whatever at present. As for his suggestion that Manderson came in by way of the window after leaving Marlowe and the car, that's right enough, I should say. I questioned the servant who swept the room next morning, and she tells me there were gravelly marks near the window, on this plain drugget that goes round the carpet. And there's a footprint in this soft new gravel just outside.' The inspector took a folding rule from his pocket and with it pointed out the traces. 'One of the patent shoes Manderson was wearing that night exactly fits that print; you'll find them,' he added, 'on the top shelf in the bedroom, near the window end, the only patents in the row. The girl who polished them in the morning picked them out for me.'

Trent bent down and studied the faint marks keenly. 'Good!' he said. 'You have covered a lot of ground, Murch, I must say. That was excellent about the whisky; you made your point finely. I felt inclined to shout "Encore!" It's a thing that I shall have to think over.'

'I thought you might have fitted it in already,' said Mr Murch. 'Come, Mr Trent, we're only at the beginning of our enquiries, but what do you say to this for a preliminary theory? There's a plan of burglary, say a couple of men in it and Martin squared. They know where the plate is, and all about the handy little bits of stuff in the drawing-room and elsewhere. They watch the house; see Manderson off to bed; Martin comes to shut the window, and leaves it ajar, accidentally on purpose. They wait till Martin goes to bed at twelve-thirty; then they just walk into the library, and begin to sample the whisky first thing. Now suppose Manderson isn't asleep, and suppose they make a noise opening the window, or however it might be. He hears it; thinks of burglars; gets up very quietly to see if anything's wrong; creeps down on them, perhaps, just as they're getting ready for work. They cut and run; he chases them down to the shed, and collars one; there's a fight; one of them loses his temper and his head, and makes a swinging job of it. Now, Mr Trent, pick that to pieces.'

'Very well,' said Trent; 'just to oblige you, Murch, especially as I know you don't believe a word of it. First: no traces of any kind left by your burglar or burglars, and the window found fastened in the morning, according to Martin. Not much force in that, I allow. Next: nobody in the house hears anything of this stampede through the library, nor hears any shout from Manderson either inside the house or outside. Next: Manderson goes down without a word to anybody, though Bunner and Martin are both at hand. Next: did you ever hear in your long experience of a householder getting up in the night to pounce on burglars, who dressed himself fully, with underclothing, shirt, collar and tie, trousers, waistcoat and coat, socks and hard leather shoes; and who gave the finishing touches to a somewhat dandified toilet by doing his hair, and putting on his watch and chain? Personally, I call that over-dressing the part. The only decorative detail he seems to have forgotten is his teeth.'

The inspector leaned forward thinking, his large hands clasped before him. 'No,' he said at last. 'Of course there's no help in that theory. I rather expect we have some way to go before we find out why a man gets up before the servants are awake, dresses himself fully, and is murdered within sight of his house early enough to be cold and stiff by ten in the morning.'

Trent shook his head. 'We can't build anything on that last consideration. I've gone into the subject with people who know. I shouldn't wonder,' he added, 'if the traditional notions about loss of temperature and rigour after death had occasionally brought an innocent man to the gallows, or near it. Dr Stock has them all, I feel sure; most general practitioners of the older generation have. That Dr Stock will make an ass of himself at the inquest, is almost as certain as that tomorrow's sun will rise. I've seen him. He will say the body must have been dead about so long, because of the degree of coldness and *rigor mortis*. I can see him nosing it all out in some textbook that was out of date when he was a student. Listen, Murch, and I will tell you some facts which will be a great hindrance to you in your professional career. There are many things that may hasten or retard the cooling of the body. This one was lying in the long dewy grass on the shady side of the shed. As for rigidity, if Manderson died in a struggle, or labouring under sudden emotion, his corpse might stiffen practically instantaneously; there are dozens of cases noted, particularly in cases of injury to the skull, like this one. On the other hand, the stiffening might not have begun until eight or ten hours after death. You can't hang anybody on *rigor mortis* nowadays, inspector, much as you may resent the limitation. No, what we *can* say is this. If he had been shot after the hour at which the world begins to get up and go about its business, it would have been heard, and very likely seen too. In fact, we must reason, to begin with, at any rate, on the assumption that he wasn't shot at a time when people might be awake; it isn't done in these parts. Put that time at 6.30 a.m. Manderson went up to bed at 11 p.m., and Martin sat up till 12.30. Assuming that he went to sleep at once on turning in, that leaves us something like six hours for the crime to be committed in; and that is a long time. But whenever it took place, I wish you would suggest a reason why Manderson, who was a fairly late riser, was up and dressed at or before 6.30; and why neither Martin, who sleeps lightly, nor Bunner, nor his wife heard him moving about, or letting himself out of the house. He must have been careful. He must have crept about like a cat. Do you feel as I do, Murch, about all this; that it is very, very strange and baffling?'

'That's how it looks,' agreed the inspector.

'And now,' said Trent, rising to his feet, 'I'll leave you to your meditations, and take a look at the bedrooms. Perhaps the explanation of all this will suddenly burst upon you while I am poking about up there. But,' concluded Trent in a voice of sudden exasperation, turning round in the doorway, 'if you can tell me at any time, how under the sun a man who put on all those clothes could forget to put in his teeth, you may kick me from here to the nearest lunatic asylum, and hand me over as an incipient dement.'

CHAPTER V

Poking About

THERE are moments in life, as one might think, when that which is within us, busy about its secret affair, lets escape into consciousness some hint of a fortunate thing ordained. Who does not know what it is to feel at times a wave of unaccountable persuasion that it is about to go well with him?—not the feverish confidence of men in danger of a blow from fate, not the persistent illusion of the optimist, but an unsought conviction, springing up like a bird from the heather, that success is at hand in some great or little thing. The general suddenly knows at dawn that the day will bring him victory; the man on the green suddenly knows that he will put down the long putt. As Trent mounted the stairway outside the library door he seemed to rise into certainty of achievement.

A host of guesses and inferences swarmed apparently unsorted through his mind; a few secret observations that he had made, and which he felt must have significance, still stood unrelated to any plausible theory of the crime; yet as he went up he seemed to know indubitably that light was going to appear.

The bedrooms lay on either side of a broad carpeted passage, lighted by a tall end window. It went the length of the house until it ran at right angles into a narrower passage, out of which the servants' rooms opened. Martin's room was the exception: it opened out of a small landing half-way to the upper floor. As Trent passed it he glanced within. A little square room, clean and commonplace. In going up the rest of the stairway he stepped with elaborate precaution against noise, hugging the wall closely and placing each foot with care; but a series of very audible creaks marked his passage.

He knew that Manderson's room was the first on the right hand when the bedroom floor was reached, and he went to it at once. He tried the latch and the lock, which worked normally, and examined the wards of the key. Then he turned to the room.

It was a small apartment, strangely bare. The plutocrat's toilet appointments were of the simplest. All remained just as it had been on the morning of the ghastly discovery in the grounds. The sheets and blankets of the unmade bed lay tumbled over a narrow wooden bedstead, and the sun shone brightly through the window upon them. It gleamed, too, upon the gold parts of the delicate work of dentistry that lay in water in a shallow bowl of glass placed on a small, plain table by the bedside. On this also stood a wrought-iron candlestick. Some clothing lay untidily over one of the two rush-bottomed chairs. Various objects on the top of a chest of drawers, which had been used as a dressing-table, lay in such disorder as a hurried man might make. Trent looked them over with a questing eye. He noted also that the occupant of the room had neither washed nor shaved. With his finger he turned over the dental plate in the bowl, and frowned again at its incomprehensible presence.

The emptiness and disarray of the little room, flooded by the sunbeams, were producing in Trent a sense of gruesomeness. His fancy called up a picture of a haggard man dressing himself in careful silence by the first light of dawn, glancing constantly at the inner door behind which his wife slept, his eyes full of some terror.

Trent shivered, and to fix his mind again on actualities, opened two tall cupboards in the wall on either side of the bed. They contained clothing, a large choice of which had evidently been one of the very few conditions of comfort for the man who had slept there.

In the matter of shoes, also, Manderson had allowed himself the advantage of wealth. An extraordinary number of these, treed and carefully kept, was ranged on two long low shelves against the wall. No boots were among them. Trent, himself an amateur of good shoe-leather, now turned to these, and glanced over the collection with an appreciative eye. It was to be seen that Manderson had been inclined to pride himself on a rather small and well-formed foot. The shoes were of a distinctive shape, narrow and round-toed, beautifully made; all were evidently from the same last.

Suddenly his eyes narrowed themselves over a pair of patent-leather shoes on the upper shelf.

These were the shoes of which the inspector had already described the position to him; the shoes worn by Manderson the night before his death. They were a well-worn pair, he saw at once; he saw, too, that they had been very recently polished. Something about the uppers of these shoes had seized his attention. He bent lower and frowned over them, comparing what he saw with the appearance of the neighbouring shoes. Then he took them up and examined the line of junction of the uppers with the soles.

As he did this, Trent began unconsciously to whistle faintly, and with great precision, an air which Inspector Murch, if he had been present, would have recognized.

Most men who have the habit of self-control have also some involuntary trick which tells those who know them that they are suppressing excitement. The inspector had noted that when Trent had picked up a strong scent he whistled faintly a certain melodious passage; though the inspector could not have told you that it was in fact the opening movement of Mendelssohn's *Lied ohne Worter* in A Major.

He turned the shoes over, made some measurements with a marked tape, and looked minutely at the bottoms. On each, in the angle between the heel and the instep, he detected a faint trace of red gravel.

Trent placed the shoes on the floor, and walked with his hands behind him to the window, out of which, still faintly whistling, he gazed with eyes that saw nothing. Once his lips opened to emit mechanically the Englishman's expletive of sudden enlightenment. At length he turned to the shelves again, and swiftly but carefully examined every one of the shoes there.

This done, he took up the garments from the chair, looked them over closely and replaced them. He turned to the wardrobe cupboards again, and hunted through them carefully. The litter on the dressing-table now engaged his attention for the second time. Then he sat down on the empty chair, took his head in his hands, and remained in that attitude, staring at the carpet, for some minutes. He rose at last and opened the inner door leading to Mrs Manderson's room.

It was evident at a glance that the big room had been hurriedly put down from its place as the lady's bower. All the array of objects that belong to a woman's dressing-table had been removed; on bed and chairs and smaller tables there were no garments or hats, bags or boxes; no trace remained of the obstinate conspiracy of gloves and veils, handkerchiefs and ribbons, to break the captivity of the drawer. The room was like an unoccupied guest-chamber. Yet in every detail of furniture and decoration it spoke of an unconventional but exacting taste. Trent, as his expert eye noted the various perfection of colour and form amid which the ill-mated lady dreamed her dreams and thought her loneliest thoughts, knew that she had at least the resources of an artistic nature. His interest in this unknown personality grew stronger; and his brows came down heavily as he thought of the burdens laid upon it, and of the deed of which the history was now shaping itself with more and more of substance before his busy mind.

He went first to the tall French window in the middle of the wall that faced the door, and opening it, stepped out upon a small balcony with an iron railing. He looked down on a broad stretch of lawn that began immediately beneath him, separated from the house-wall only by a narrow flower-bed, and stretched away, with an abrupt dip at the farther end, toward the orchard. The other window opened with a sash above the garden-entrance of the library. In the farther inside corner of the room was a second door giving upon the passage; the door by which the maid was wont to come in, and her mistress to go out, in the morning.

Trent, seated on the bed, quickly sketched in his notebook a plan of the room and its neighbour. The bed stood in the angle between the communicating-door and the sash-window, its head against the wall dividing the room from Manderson's. Trent stared at the pillows; then he lay down with deliberation on the bed and looked through the open door into the adjoining room.

This observation taken, he rose again and proceeded to note on his plan that on either side of the bed was a small table with a cover. Upon that furthest from the door was a graceful electric-lamp standard of copper connected by a free wire with the wall. Trent looked at it thoughtfully, then at the switches connected with the other lights in the room. They were, as usual, on the wall just within the door, and some way out of his reach as he sat on the bed. He rose, and satisfied himself that the lights were all in order. Then he turned on his heel, walked quickly into Manderson's room, and rang the bell.

'I want your help again, Martin,' he said, as the butler presented himself, upright and impassive, in the doorway. 'I want you to prevail upon Mrs Manderson's maid to grant me an interview.'

'Certainly, sir,' said Martin.

'What sort of a woman is she? Has she her wits about her?'

'She's French, sir,' replied Martin succinctly; adding after a pause: 'She has not been with us long, sir, but I have formed the impression that the young woman knows as much of the world as is good for her—since you ask me.'

'You think butter might possibly melt in her mouth, do you?' said Trent. 'Well, I am not afraid. I want to put some questions to her.'

'I will send her up immediately, sir.' The butler withdrew, and Trent wandered round the little room with his hands at his back. Sooner than he had expected, a small neat figure in black appeared quietly before him.

The lady's maid, with her large brown eyes, had taken favourable notice of Trent from a window when he had crossed the lawn, and had been hoping desperately that the resolver of mysteries (whose reputation was as great below-stairs as elsewhere) would send for her. For one thing, she felt the need to make a scene; her nerves were overwrought. But her scenes were at a discount with the other domestics, and as for Mr Murch, he had chilled her into self-control with his official manner. Trent, her glimpse of him had told her, had not the air of a policeman, and at a distance he had appeared *sympathique*.

As she entered the room, however, instinct decided for her that any approach to coquetry would be a mistake, if she sought to make a good impression at the beginning. It was with an air of amiable candour, then, that she said, 'Monsieur desire to speak with me.' She added helpfully, 'I am called Célestine.'

'Naturally,' said Trent with businesslike calm. 'Now what I want you to tell me, Célestine, is this. When you took tea to your mistress yesterday morning at seven o'clock, was the door between the two bedrooms—this door here—open?'

Célestine became intensely animated in an instant. 'Oh yes!' she said, using her favourite English idiom. 'The door was open as always, monsieur, and I shut it as always. But it is necessary to explain. Listen! When I enter the room of madame from the other door in there—ah! but if monsieur will give himself the pain to enter the other room, all explains itself.' She tripped across to the door, and urged Trent before her into the larger bedroom with a hand on his arm. 'See! I enter the room with the tea like this. I approach the bed. Before I come quite near the bed, here is the door to my right hand—open always—so! But monsieur can perceive that I see nothing in the room of Monsieur Manderson. The door opens to the bed, not to me who approach from down there. I shut it without seeing in. It is the order. Yesterday it was as ordinary. I see nothing of the next room. Madame sleep like an angel—she see nothing. I shut the door. I place the *plateau*—I open the curtains—I prepare the toilette—I retire— voilà!' Célestine paused for breath and spread her hands abroad.

Trent, who had followed her movements and gesticulations with deepening gravity, nodded his head. 'I see exactly how it was now,' he said. 'Thank you, Célestine. So Mr Manderson was supposed to be still in his room while your mistress was getting up, and dressing, and having breakfast in her boudoir?'

'Oui, monsieur.'

'Nobody missed him, in fact,' remarked Trent. 'Well, Célestine, I am very much obliged to you.' He reopened the door to the outer bedroom.

'It is nothing, monsieur,' said Célestine, as she crossed the small room. 'I hope that monsieur will catch the assassin of Monsieur Manderson. But I not regret him too much,' she added with sudden and amazing violence, turning round with her hand on the knob of the outer door. She set her teeth with an audible sound, and the colour rose in her small dark face. English departed from her. 'Je ne le regrette pas du tout, du tout!' she cried with a flood of words. 'Madame—ah! je me jetterais au feu pour madame—une femme si charmante, si adorable! Mais un homme comme monsieur—maussade, boudeur, impassible! Ah, non!—de ma vie! J'en avais par-dessus la tête, de monsieur! Ah! vrai! Est-ce insupportable, tout de même, qu'il existe des types comme ça? Je vous jure que—'

'Finissez ce chahut, Célestine!' Trent broke in sharply. Célestine's tirade had brought back the memory of his student days with a rush. 'En voilà une scène! C'est rasant, vous savez. Faut rentrer ça, mademoiselle. Du reste, c'est bien imprudent, croyez-moi. Hang it! have some common sense! If the inspector downstairs heard you saying that kind of thing, you would get into trouble. And don't wave your fists about so much; you might hit something. You seem,' he went on more pleasantly, as Célestine grew calmer under his authoritative eye, 'to be even more glad than other people that Mr Manderson is out of the way. I could almost suspect, Célestine, that Mr Manderson did not take as much notice of you as you thought necessary and right.'

'A peine s'il m'avait regardé!' Célestine answered simply.

'Ça, c'est un comble!' observed Trent. 'You are a nice young woman for a small tea-party, I don't think. A star upon your birthday burned, whose fierce, serene, red, pulseless planet never yearned in heaven, Célestine. Mademoiselle, I am busy. Bon jour. You certainly are a beauty!'

Célestine took this as a scarcely expected compliment. The surprise restored her balance. With a sudden flash of her eyes and teeth at Trent over her shoulder, the lady's maid opened the door and swiftly disappeared.

Trent, left alone in the little bedroom, relieved his mind with two forcible descriptive terms in Célestine's language, and turned to his problem. He took the pair of shoes which he had already examined, and placed them on one of the two chairs in the room, then seated himself on the other opposite to this. With his hands in his pockets he sat with eyes fixed upon those two dumb witnesses. Now and then he whistled, almost inaudibly, a few bars. It was very still in the room. A subdued twittering came from the trees through the open window. From time to time a breeze rustled in the leaves of the thick creeper about the sill. But the man in the room, his face grown hard and sombre now with his thoughts, never moved.

So he sat for the space of half an hour. Then he rose quickly to his feet. He replaced the shoes on their shelf with care, and stepped out upon the landing.

Two bedroom doors faced him on the other side of the passage. He opened that which was immediately opposite, and entered a bedroom by no means austerely tidy. Some sticks and fishing-rods stood confusedly in one corner, a pile of books in another. The housemaid's hand had failed to give a look of order to the jumble of heterogeneous objects left on the dressing-table and on the mantelshelf—pipes, penknives, pencils, keys, golf-balls, old letters, photographs, small boxes, tins, and bottles. Two fine etchings and some water-colour sketches hung on the walls; leaning against the end of the wardrobe, unhung, were a few framed engravings. A row of shoes and boots was ranged beneath the window. Trent crossed the room and studied them intently; then he measured some of them with his tape, whistling very softly. This done, he sat on the side of the bed, and his eyes roamed gloomily about the room.

The photographs on the mantelshelf attracted him presently. He rose and examined one representing Marlowe and Manderson on horseback. Two others were views of famous peaks in the Alps. There was a faded print of three youths—one of them unmistakably his acquaintance of the haggard blue eyes—clothed in tatterdemalion soldier's gear of the sixteenth century. Another was a portrait of a majestic old lady, slightly resembling Marlowe. Trent, mechanically taking a cigarette from an open box on the mantel-shelf, lit it and stared at the photographs. Next he turned his attention to a flat leathern case that lay by the cigarette-box.

It opened easily. A small and light revolver, of beautiful workman-ship, was disclosed, with a score or so of loose cartridges. On the stock were engraved the initials 'J. M.'

A step was heard on the stairs, and as Trent opened the breech and peered into the barrel of the weapon, Inspector Murch appeared at the open door of the room. 'I was wondering—' he began; then stopped as he saw what the other was about. His intelligent eyes opened slightly. 'Whose is the revolver, Mr Trent?' he asked in a conversational tone.

'Evidently it belongs to the occupant of the room, Mr Marlowe,' replied Trent with similar lightness, pointing to the initials. 'I found this lying about on the mantelpiece. It seems a handy little pistol to me, and it has been very carefully cleaned, I should say, since the last time it was used. But I know little about firearms.'

'Well, I know a good deal,' rejoined the inspector quietly, taking the revolver from Trent's outstretched hand. 'It's a bit of a speciality with me, is firearms, as I think you know, Mr Trent. But it don't require an expert to tell one thing.' He replaced the revolver in its case on the mantel-shelf, took out one of the cartridges, and laid it on the spacious palm of one hand; then, taking a small object from his waistcoat pocket, he laid it beside the cartridge. It was a little leaden bullet, slightly battered about the nose, and having upon it some bright new scratches.

'Is that *the* one?' Trent murmured as he bent over the inspector's hand.

'That's him,' replied Mr Murch. 'Lodged in the bone at the back of the skull. Dr Stock got it out within the last hour, and handed it to the local officer, who has just sent it on to me. These bright scratches you see were made by the doctor's instruments. These other marks were made by the rifling of the barrel—a barrel like this one.' He tapped the revolver. 'Same make, same calibre. There is no other that marks the bullet just like this.'

With the pistol in its case between them, Trent and the inspector looked into each other's eyes for some moments. Trent was the first to speak. 'This mystery is all wrong,' he observed. 'It is insanity. The symptoms of mania are very marked. Let us see how we stand. We were not in any doubt, I believe, about Manderson having dispatched Marlowe in the car to Southampton, or about Marlowe having gone, returning late last night, many hours after the murder was committed.'

'There *is* no doubt whatever about all that,' said Mr Murch, with a slight emphasis on the verb.

'And now,' pursued Trent, 'we are invited by this polished and insinuating firearm to believe the following line of propositions: that Marlowe never went to Southampton; that he returned to the house in the night; that he somehow, without waking Mrs Manderson or anybody else, got Manderson to get up, dress himself, and go out into the grounds; that he then and there shot the said Manderson with his incriminating pistol; that he carefully cleaned the said pistol, returned to the house and, again without disturbing any one, replaced it in its case in a favourable position to be found by the officers of the law; that he then withdrew and spent the rest of the day in hiding—*with* a large motor car; and that he turned up, feigning ignorance of the whole affair, at—what time was it?'

'A little after 9 p.m.' The inspector still stared moodily at Trent. 'As you say, Mr Trent, that is the first theory suggested by this find, and it seems wild enough—at least it would do if it didn't fall to pieces at the very start. When the murder was done Marlowe must have been fifty to a hundred miles away. He *did* go to Southampton.'

'How do you know?'

'I questioned him last night, and took down his story. He arrived in Southampton about 6.30 on the Monday morning.'

'Come off!' exclaimed Trent bitterly. 'What do I care about his story? What do you care about his story? I want to know how you *know* he went to Southampton.'

Mr Murch chuckled. 'I thought I should take a rise out of you, Mr Trent,' he said. 'Well, there's no harm in telling you. After I arrived yesterday evening, as soon as I had got the outlines of the story from Mrs Manderson and the servants, the first thing I did was to go to the telegraph office and wire to our people in Southampton. Manderson had told his wife when he went to bed that he had changed his mind, and sent Marlowe to Southampton to get some important information from some one who was crossing by the next day's boat. It seemed right enough, but, you see, Marlowe was the only one of the household who wasn't under my hand, so to speak. He didn't return in the car until later in the evening; so before thinking the matter out any further, I wired to Southampton making certain enquiries. Early this morning I got this reply.' He handed a series of telegraph slips to Trent, who read:

Person answering description in motor answering description arrived Bedford Hotel here 6.30 this morning gave name Marlowe left car hotel garage told attendant car belonged Manderson had bath and breakfast went out heard of later at docks enquiring for passenger name Harris on Havre boat enquired repeatedly until boat left at noon next heard of at hotel where he lunched about 1.15 left soon afterwards in car company's agents inform berth was booked name Harris last week but Harris did not travel by boat   Burke Inspector.

'Simple and satisfactory,' observed Mr Murch as Trent, after twice reading the message, returned it to him. 'His own story corroborated in every particular. He told me he hung about the dock for half an hour or so on the chance of Harris turning up late, then strolled back, lunched, and decided to return at once. He sent a wire to Manderson—"Harris not turned up missed boat returning Marlowe," which was duly delivered here in the afternoon, and placed among the dead man's letters. He motored back at a good rate, and arrived dog-tired. When he heard of Manderson's death from Martin, he nearly fainted. What with that and the being without sleep for so long, he was rather a wreck when I came to interview him last night; but he was perfectly coherent.'

Trent picked up the revolver and twirled the cylinder idly for a few moments. 'It was unlucky for Manderson that Marlowe left his pistol and cartridges about so carelessly,' he remarked at length, as he put it back in the case. 'It was throwing temptation in somebody's way, don't you think?'

Mr Murch shook his head. 'There isn't really much to lay hold of about the revolver, when you come to think. That particular make of revolver is common enough in England. It was introduced from the

States. Half the people who buy a revolver today for self-defence or mischief provide themselves with that make, of that calibre. It is very reliable, and easily carried in the hip-pocket. There must be thousands of them in the possession of crooks and honest men. For instance,' continued the inspector with an air of unconcern, 'Manderson himself had one, the double of this. I found it in one of the top drawers of the desk downstairs, and it's in my overcoat pocket now.'

'Aha! so you were going to keep that little detail to yourself.'

'I was,' said the inspector; 'but as you've found one revolver, you may as well know about the other. As I say, neither of them may do us any good. The people in the house—'

Both men started, and the inspector checked his speech abruptly, as the half-closed door of the bedroom was slowly pushed open, and a man stood in the doorway. His eyes turned from the pistol in its open case to the faces of Trent and the inspector. They, who had not heard a sound to herald this entrance, simultaneously looked at his long, narrow feet. He wore rubber-soled tennis shoes.

'You must be Mr Bunner,' said Trent.

CHAPTER VI

Mr Bunner on the Case

'CALVIN C. BUNNER, at your service,' amended the newcomer, with a touch of punctilio, as he removed an unlighted cigar from his mouth. He was used to finding Englishmen slow and ceremonious with strangers, and Trent's quick remark plainly disconcerted him a little. 'You are Mr Trent, I expect,' he went on. 'Mrs Manderson was telling me a while ago. Captain, good-morning.' Mr Murch acknowledged the outlandish greeting with a nod. 'I was coming up to my room, and I heard a strange voice in here, so I thought I would take a look in.' Mr Bunner laughed easily. 'You thought I might have been eavesdropping, perhaps,' he said. 'No, sir; I heard a word or two about a pistol—this one, I guess—and that's all.'

Mr Bunner was a thin, rather short young man with a shaven, pale, bony, almost girlish face, and large, dark, intelligent eyes. His waving dark hair was parted in the middle. His lips, usually occupied with a cigar, in its absence were always half open with a curious expression as of permanent eagerness. By smoking or chewing a cigar this expression was banished, and Mr Bunner then looked the consummately cool and sagacious Yankee that he was.

Born in Connecticut, he had gone into a broker's office on leaving college, and had attracted the notice of Manderson, whose business with his firm he had often handled. The Colossus had watched him for some time, and at length offered him the post of private secretary. Mr Bunner was a pattern business man, trustworthy, long-headed, methodical, and accurate. Manderson could have found many men with those virtues; but he engaged Mr Bunner because he was also swift and secret, and had besides a singular natural instinct in regard to the movements of the stock market.

Trent and the American measured one another coolly with their eyes. Both appeared satisfied with what they saw. 'I was having it explained to me,' said Trent pleasantly, 'that my discovery of a pistol that might have shot Manderson does not amount to very much. I am told it is a favourite weapon among your people, and has become quite popular over here.'

Mr Bunner stretched out a bony hand and took the pistol from its case. 'Yes, sir,' he said, handling it with an air of familiarity; 'the captain is right. This is what we call out home a Little Arthur, and I dare say there are duplicates of it in a hundred thousand hip-pockets this minute. I consider it too light in the hand myself,' Mr Bunner went on, mechanically feeling under the tail of his jacket, and producing an ugly looking weapon. 'Feel of that, now, Mr Trent—it's loaded, by the way. Now this Little Arthur—Marlowe bought it just before we came over this year to please the old man. Manderson said it was ridiculous for a man to be without a pistol in the twentieth century. So he went out and bought what they offered him, I guess— never consulted me. Not but what it's a good gun,' Mr Bunner conceded, squinting along the sights. 'Marlowe was poor with it at first, but I've coached him some in the last month or so, and he's practised until he is pretty good. But he never could get the habit of carrying it around. Why, it's as natural to me as wearing my pants. I have carried one for some years now, because there was always likely to be somebody laying for Manderson. And now,' Mr Bunner concluded sadly, 'they got him when I wasn't around. Well, gentlemen, you must excuse me. I am going into Bishopsbridge. There is a lot to do these days, and I have to send off a bunch of cables big enough to choke a cow.'

'I must be off too,' said Trent. 'I have an appointment at the "Three Tuns" inn.'

Let me give you a lift in the automobile,' said Mr Bunner cordially. 'I go right by that joint. Say, cap., are you coming my way too? No? Then come along, Mr Trent, and help me get out the car. The chauffeur is out of action, and we have to do 'most everything ourselves except clean the dirt off her.'

Still tirelessly talking in his measured drawl, Mr Bunner led Trent downstairs and through the house to the garage at the back. It stood at a little distance from the house, and made a cool retreat from the blaze of the midday sun.

Mr Bunner seemed to be in no hurry to get out the car. He offered Trent a cigar, which was accepted, and for the first time lit his own. Then he seated himself on the footboard of the car, his thin hands clasped between his knees, and looked keenly at the other.

'See here, Mr Trent,' he said, after a few moments. 'There are some things I can tell you that may be useful to you. I know your record. You are a smart man, and I like dealing with smart men. I don't know if I have that detective sized up right, but he strikes me as a mutt. I would answer any questions he had the gumption to ask me—I have done so, in fact—but I don't feel encouraged to give him any notions of mine without his asking. See?'

Trent nodded. 'That is a feeling many people have in the presence of our police,' he said. 'It's the official manner, I suppose. But let me tell you, Murch is anything but what you think. He is one of the shrewdest officers in Europe. He is not very quick with his mind, but he is very sure. And his experience is immense. My forte is imagination, but I assure you in police work experience outweighs it by a great deal.'

'Outweigh nothing!' replied Mr Bunner crisply. 'This is no ordinary case, Mr Trent. I will tell you one reason why. I believe the old man knew there was something coming to him. Another thing: I believe it was something he thought he couldn't dodge.'

Trent pulled a crate opposite to Mr Bunner's place on the footboard and seated himself. 'This sounds like business,' he said. 'Tell me your ideas.'

'I say what I do because of the change in the old man's manner this last few weeks. I dare say you have heard, Mr Trent, that he was a man who always kept himself well in hand. That was so. I have always considered him the coolest and hardest head in business. That man's calm was just deadly—I never saw anything to beat it. And I knew Manderson as nobody else did. I was with him in the work he really lived for. I guess I knew him a heap better than his wife did, poor woman. I knew him better than Marlowe could—he never saw Manderson in his office when there was a big thing on. I knew him better than any of his friends.'

'Had he any friends?' interjected Trent.

Mr Bunner glanced at him sharply. 'Somebody has been putting you next, I see that,' he remarked. 'No: properly speaking, I should say not. He had many acquaintances among the big men, people he saw, most every day; they would even go yachting or hunting together. But I don't believe there ever was a man that Manderson opened a corner of his heart to. But what I was going to say was this. Some months ago the old man began to get like I never knew him before—gloomy and sullen, just as if he was everlastingly brooding over something bad, something that he couldn't fix. This went on without any break; it was the same down town as it was up home, he acted just as if there was something lying heavy on his mind. But it wasn't until a few weeks back that his self-restraint began to go; and let me tell you this, Mr Trent'—the American laid his bony claw on the other's knee—'I'm the only man that knows it. With every one else he would be just morose and dull; but when he was alone with me in his office, or anywhere where we would be working together, if the least little thing went wrong, by George! he would fly off the handle to beat the Dutch. In this library here I have seen him open a letter with something that didn't just suit him in it, and he would rip around and carry on like an Indian, saying he wished he had the man that wrote it here, he wouldn't do a thing to him, and so on, till it was just pitiful. I never saw such a change. And here's another thing. For a week before he died Manderson neglected his work, for the first time in my experience. He wouldn't answer a letter or a cable, though things looked like going all to pieces over there. I supposed that this anxiety of his, whatever it was, had got on to his nerves till they were worn out. Once I advised him to see a doctor, and he told me to go to hell. But nobody saw this side of him but me. If he was having one of these rages in the library here, for example, and Mrs Manderson would come into the room, he would be all calm and cold again in an instant.'

'And you put this down to some secret anxiety, a fear that some-body had designs on his life?' asked Trent.

The American nodded.

'I suppose,' Trent resumed, 'you had considered the idea of there being something wrong with his mind—a break-down from overstrain, say. That is the first thought that your account suggests to me. Besides, it is what is always happening to your big business men in America, isn't it? That is the impression one gets from the newspapers.'

'Don't let them slip you any of that bunk,' said Mr Bunner earnestly. 'It's only the ones who have got rich too quick, and can't make good, who go crazy. Think of all our really big men—the men anywhere near Manderson's size: did you ever hear of any one of them losing his senses? They don't do it—believe *me*. I know they say every man has his loco point,' Mr Bunner added reflectively, 'but that doesn't mean genuine, sure-enough craziness; it just means some personal eccentricity in a man . . . like hating cats . . . or my own weakness of not being able to touch any kind of fish-food.'

'Well, what was Manderson's?'

'He was full of them—the old man. There was his objection to all the unnecessary fuss and luxury that wealthy people don't kick at much, as a general rule. He didn't have any use for expensive trifles and ornaments. He wouldn't have anybody do little things for him; he hated to have servants tag around after him unless he wanted them. And although Manderson was as careful about his clothes as any man I ever knew, and his shoes—well, sir, the amount of money he spent on shoes was sinful—in spite of that, I tell you, he never had a valet. He never liked to have anybody touch him. All his life nobody ever shaved him.'

'I've heard something of that,' Trent remarked. 'Why was it, do you think?'

'Well,' Mr Bunner answered slowly, 'it was the Manderson habit of mind, I guess; a sort of temper of general suspicion and jealousy. They say his father and grandfather were just the same. . . . Like a dog with a bone, you know, acting as if all the rest of creation was laying for a chance to steal it. He didn't really *think* the barber would start in to saw his head off; he just felt there was a possibility that he *might*, and he was taking no risks. Then again in business he was always convinced that somebody else was after his bone—which was true enough a good deal of the time; but not all the time. The consequence of that was that the old man was the most cautious and secret worker in the world of finance; and that had a lot to do with his success, too. . . . But that doesn't amount to being a lunatic, Mr Trent; not by a long way. You ask me if Manderson was losing his mind before he died. I say I believe he was just worn out with worrying over something, and was losing his nerve.'

Trent smoked thoughtfully. He wondered how much Mr Bunner knew of the domestic difficulty in his chief's household, and decided to put out a feeler. 'I understood that he had trouble with his wife.'

'Sure,' replied Mr Bunner. 'But do you suppose a thing like that was going to upset Sig Manderson that way? No, sir! He was a sight too big a man to be all broken up by any worry of that kind.'

Trent looked half-incredulously into the eyes of the young man. But behind all their shrewdness and intensity he saw a massive

innocence. Mr Bunner really believed a serious breach between husband and wife to be a minor source of trouble for a big man.

'What *was* the trouble between them, anyhow?' Trent enquired.

'You can search me,' Mr Bunner replied briefly. He puffed at his
cigar. 'Marlowe and I have often talked about it, and we could never
make out a solution. I had a notion at first,' said Mr Bunner in a lower
voice, leaning forward, 'that the old man was disappointed and vexed
because he had expected a child; but Marlowe told me that the
disappointment on that score was the other way around, likely as not.
His idea was all right, I guess; he gathered it from something said by
Mrs Manderson's French maid.'

Trent looked up at him quickly. 'Célestine!' he said; and his
thought was, 'So that was what she was getting at!'

Mr Bunner misunderstood his glance. 'Don't you think I'm giving
a man away, Mr Trent,' he said. 'Marlowe isn't that kind. Célestine
just took a fancy to him because he talks French like a native, and she
would always be holding him up for a gossip. French servants are
quite unlike English that way. And servant or no servant,' added Mr
Bunner with emphasis, 'I don't see how a woman could mention such
a subject to a man. But the French beat me.' He shook his head
slowly.

'But to come back to what you were telling me just now,' Trent
said. 'You believe that Manderson was going in terror of his life for
some time. Who should threaten it? I am quite in the dark.'

'Terror—I don't know,' replied Mr Bunner meditatively. 'Anxiety,
if you like. Or suspense—that's rather my idea of it. The old man was
hard to terrify, anyway; and more than that, he wasn't taking any
precautions—he was actually avoiding them. It looked more like he
was asking for a quick finish—supposing there's any truth in my idea.
Why, he would sit in that library window, nights, looking out into the
dark, with his white shirt just a target for anybody's gun. As for who
should threaten his life—well, sir,' said Mr Bunner with a faint smile,
'it's certain you have not lived in the States. To take the Pennsylvania
coal hold-up alone, there were thirty thousand men, with women and
children to keep, who would have jumped at the chance of drilling a
hole through the man who fixed it so that they must starve or give in
to his terms. Thirty thousand of the toughest aliens in the country,
Mr Trent. There's a type of desperado you find in that kind of push
who has been known to lay for a man for years, and kill him when he

had forgotten what he did. They have been known to dynamite a man in Idaho who had done them dirt in New Jersey ten years before. Do you suppose the Atlantic is going to stop them? . . . It takes some sand, I tell you, to be a big business man in our country. No, sir: the old man knew—had always known—that there was a whole crowd of dangerous men scattered up and down the States who had it in for him. My belief is that he had somehow got to know that some of them were definitely after him at last. What licks me altogether is why he should have just laid himself open to them the way he did—why he never tried to dodge, but walked right down into the garden yesterday morning to be shot at.'

Mr Bunner ceased to speak, and for a little while both men sat with wrinkled brows, faint blue vapours rising from their cigars. Then Trent rose. 'Your theory is quite fresh to me,' he said. 'It's perfectly rational, and it's only a question of whether it fits all the facts. I mustn't give away what I'm doing for my newspaper, Mr Bunner, but I will say this: I have already satisfied myself that this was a premeditated crime, and an extraordinarily cunning one at that. I'm deeply obliged to you. We must talk it over again.' He looked at his watch. 'I have been expected for some time by my friend. Shall we make a move?'

'Two o'clock,' said Mr Bunner, consulting his own, as he got up from the foot-board. 'Ten a.m. in little old New York. You don't know Wall Street, Mr Trent. Let's you and I hope we never see anything nearer hell than what's loose in the Street this minute.'

CHAPTER VII

The Lady in Black

THE sea broke raging upon the foot of the cliff under a good breeze; the sun flooded the land with life from a dappled blue sky. In this perfection of English weather Trent, who had slept ill, went down before eight o'clock to a pool among the rocks, the direction of which had been given him, and dived deep into clear water. Between vast grey boulders he swam out to the tossing open, forced himself some little way against a coast-wise current, and then returned to his refuge battered and refreshed. Ten minutes later he was scaling the cliff again, and his mind, cleared for the moment of a heavy disgust for the affair he had in hand, was turning over his plans for the morning.

It was the day of the inquest, the day after his arrival in the place. He had carried matters not much further after parting with the American on the road to Bishopsbridge. In the afternoon he had walked from the inn into the town, accompanied by Mr Cupples, and had there made certain purchases at a chemist's shop, conferred privately for some time with a photographer, sent off a reply-paid telegram, and made an enquiry at the telephone exchange. He had said but little about the case to Mr Cupples, who seemed incurious on his side, and nothing at all about the results of his investigation or the steps he was about to take. After their return from Bishopsbridge, Trent had written a long dispatch for the *Record* and sent it to be telegraphed by the proud hands of the paper's local representative. He had afterwards dined with Mr Cupples, and had spent the rest of the evening in meditative solitude on the veranda.

This morning as he scaled the cliff he told himself that he had never taken up a case he liked so little, or which absorbed him so much. The more he contemplated it in the golden sunshine of this new day, the more evil and the more challenging it appeared. All that he suspected and all that he almost knew had occupied his questing brain for hours to the exclusion of sleep; and in this glorious light and air, though washed in body and spirit by the fierce purity of the sea,

he only saw the more clearly the darkness of the guilt in which he
believed, and was more bitterly repelled by the motive at which he
guessed. But now at least his zeal was awake again, and the sense of
the hunt quickened. He would neither slacken nor spare; here need
be no compunction. In the course of the day, he hoped, his net would
be complete. He had work to do in the morning; and with very vivid
expectancy, though not much serious hope, he awaited the answer to
the telegram which he had shot into the sky, as it were, the day
before.

The path back to the hotel wound for some way along the top of
the cliff, and on nearing a spot he had marked from the sea-level,
where the face had fallen away long ago, he approached the edge and
looked down, hoping to follow with his eyes the most delicately
beautiful of all the movements of water—the wash of a light sea over
broken rock. But no rock was there. A few feet below him a broad
ledge stood out, a rough platform as large as a great room, thickly
grown with wiry grass and walled in steeply on three sides. There,
close to the verge where the cliff at last dropped sheer, a woman was
sitting, her arms about her drawn-up knees, her eyes fixed on the
trailing smoke of a distant liner, her face full of some dream.

This woman seemed to Trent, whose training had taught him to
live in his eyes, to make the most beautiful picture he had ever seen.
Her face of southern pallor, touched by the kiss of the wind with
colour on the cheek, presented to him a profile of delicate regularity
in which there was nothing hard; nevertheless the black brows bending down toward the point where they almost met gave her in repose
a look of something like severity, strangely redeemed by the open
curves of the mouth. Trent said to himself that the absurdity or
otherwise of a lover writing sonnets to his mistress's eyebrow depended after all on the quality of the eyebrow. Her nose was of the
straight and fine sort, exquisitely escaping the perdition of too much
length, which makes a conscientious mind ashamed that it cannot
help, on occasion, admiring the tip-tilted. Her hat lay pinned to the
grass beside her, and the lively breeze played with her thick dark hair,
blowing backward the two broad bandeaux that should have covered
much of her forehead, and agitating a hundred tiny curls from the
mass gathered at her nape. Everything about this lady was black, from
her shoes of suede to the hat that she had discarded; lustreless black

covered her to her bare throat. All she wore was fine and well put on. Dreamy and delicate of spirit as her looks declared her, it was very plain that she was long-practised as only a woman grown can be in dressing well, the oldest of the arts, and had her touch of primal joy in the excellence of the body that was so admirably curved now in the attitude of embraced knees. With the suggestion of French taste in her clothes, she made a very modern figure seated there, until one looked at her face and saw the glow and triumph of all vigorous beings that ever faced sun and wind and sea together in the prime of the year. One saw, too, a womanhood so unmixed and vigorous, so unconsciously sure of itself, as scarcely to be English, still less American.

Trent, who had halted only for a moment in the surprise of seeing the woman in black, had passed by on the cliff above her, perceiving and feeling as he went the things set down. At all times his keen vision and active brain took in and tasted details with an easy swiftness that was marvellous to men of slower chemistry; the need to stare, he held, was evidence of blindness. Now the feeling of beauty was awakened and exultant, and doubled the power of his sense. In these instants a picture was printed on his memory that would never pass away.

As he went by unheard on the turf the woman, still alone with her thoughts, suddenly moved. She unclasped her long hands from about her knees, stretched her limbs and body with feline grace, then slowly raised her head and extended her arms with open, curving fingers, as if to gather to her all the glory and overwhelming sanity of the morning. This was a gesture not to be mistaken: it was a gesture of freedom, the movement of a soul's resolution to be, to possess, to go forward, perhaps to enjoy.

So he saw her for an instant as he passed, and he did not turn. He knew suddenly who the woman must be, and it was as if a curtain of gloom were drawn between him and the splendour of the day.

During breakfast at the hotel Mr Cupples found Trent little inclined to talk. He excused himself on the plea of a restless night. Mr Cupples, on the other hand, was in a state of bird-like alertness. The prospect of the inquest seemed to enliven him. He entertained Trent with a disquisition upon the history of that most ancient and once busy tribunal, the coroner's court, and remarked upon the enviable freedom of its procedure from the shackles of rule and precedent. From this he passed to the case that was to come before it that morning.

'Young Bunner mentioned to me last night,' he said, 'when I went up there after dinner, the hypothesis which he puts forward in regard to the crime. A very remarkable young man, Trent. His meaning is occasionally obscure, but in my opinion he is gifted with a clear-headed knowledge of the world quite unusual in one of his apparent age. Indeed, his promotion by Manderson to the position of his principal lieutenant speaks for itself. He seems to have assumed with perfect confidence the control at this end of the wire, as he expresses it, of the complicated business situation caused by the death of his principal, and he has advised very wisely as to the steps I should take on Mabel's behalf, and the best course for her to pursue until effect has been given to the provisions of the will. I was accordingly less disposed than I might otherwise have been to regard his suggestion of an industrial vendetta as far-fetched. When I questioned him he was able to describe a number of cases in which attacks of one sort or another—too often successful—had been made upon the lives of persons who had incurred the hostility of powerful labour organiza-tions. This is a terrible time in which we live, my dear boy. There is none recorded in history, I think, in which the disproportion between the material and the moral constituents of society has been so great or so menacing to the permanence of the fabric. But nowhere, in my judgement, is the prospect so dark as it is in the United States.'

'I thought,' said Trent listlessly, 'that Puritanism was about as strong there as the money-getting craze.'

'Your remark,' answered Mr Cupples, with as near an approach to humour as was possible to him, 'is not in the nature of a testimonial to what you call Puritanism—a convenient rather than an accurate term; for I need not remind you that it was invented to describe an Anglican party which aimed at the purging of the services and ritual of their Church from certain elements repugnant to them. The sense of your observation, however, is none the less sound, and its truth is extremely well illustrated by the case of Manderson himself, who had, I believe, the virtues of purity, abstinence, and self-restraint in their strongest form. No, Trent, there are other and more worthy things among the moral constituents of which I spoke; and in our finite nature, the more we preoccupy ourselves with the bewildering complexity of external apparatus which science places in our hands, the less vigour have we left for the development of the holier purposes of humanity within us. Agricultural machinery has abolished the festival of the Harvest Home. Mechanical travel has abolished the inn, or all that was best in it. I need not multiply instances. The view I am expressing to you,' pursued Mr Cupples, placidly buttering a piece of toast, 'is regarded as fundamentally erroneous by many of those who think generally as I do about the deeper concerns of life, but I am nevertheless firmly persuaded of its truth.'

'It needs epigrammatic expression,' said Trent, rising from the table. 'If only it could be crystallized into some handy formula, like "No Popery", or "Tax the Foreigner", you would find multitudes to go to the stake for it. But you were planning to go to White Gables before the inquest, I think. You ought to be off if you are to get back to the court in time. I have something to attend to there myself, so we might walk up together. I will just go and get my camera.'

'By all means,' Mr Cupples answered; and they set off at once in the ever-growing warmth of the morning. The roof of White Gables, a surly patch of dull red against the dark trees, seemed to harmonize with Trent's mood; he felt heavy, sinister, and troubled. If a blow must fall that might strike down that creature radiant of beauty and life whom he had seen that morning, he did not wish it to come from his hand. An exaggerated chivalry had lived in Trent since the first teachings of his mother; but at this moment the horror of bruising anything so lovely was almost as much the artist's revulsion as the gentleman's. On the other hand, was the hunt to end in nothing? The quality of the affair was such that the thought of forbearance was an agony. There never was such a case; and he alone, he was confident, held the truth of it under his hand. At least, he determined, that day should show whether what he believed was a delusion. He would trample his compunction underfoot until he was quite sure that there was any call for it. That same morning he would know.

As they entered at the gate of the drive they saw Marlowe and the American standing in talk before the front door. In the shadow of the porch was the lady in black.

She saw them, and came gravely forward over the lawn, moving as Trent had known that she would move, erect and balanced, stepping lightly. When she welcomed him on Mr Cupples's presentation her eyes of golden-flecked brown observed him kindly. In her pale com-

posure, worn as the mask of distress, there was no trace of the emo-
tion that had seemed a halo about her head on the ledge of the cliff.
She spoke the appropriate commonplace in a low and even voice.
After a few words to Mr Cupples she turned her eyes on Trent again.

'I hope you will succeed,' she said earnestly. 'Do you think you will
succeed?'

He made his mind up as the words left her lips. He said, 'I believe
I shall do so, Mrs Manderson. When I have the case sufficiently
complete I shall ask you to let me see you and tell you about it. It may
be necessary to consult you before the facts are published.'

She looked puzzled, and distress showed for an instant in her eyes.
'If it is necessary, of course you shall do so,' she said.

On the brink of his next speech Trent hesitated. He remembered
that the lady had not wished to repeat to him the story already given
to the inspector—or to be questioned at all. He was not unconscious
that he desired to hear her voice and watch her face a little longer, if
it might be; but the matter he had to mention really troubled his
mind, it was a queer thing that fitted nowhere into the pattern within
whose corners he had by this time brought the other queer things in
the case. It was very possible that she could explain it away in a
breath; it was unlikely that any one else could. He summoned his
resolution.

'You have been so kind,' he said, 'in allowing me access to the
house and every opportunity of studying the case, that I am going to
ask leave to put a question or two to yourself—nothing that you
would rather not answer, I think. May I?'

She glanced at him wearily. 'It would be stupid of me to refuse. Ask
your questions, Mr Trent.'

'It's only this,' said Trent hurriedly. 'We know that your husband
lately drew an unusually large sum of ready money from his London
bankers, and was keeping it here. It is here now, in fact. Have you any
idea why he should have done that?'

She opened her eyes in astonishment. 'I cannot imagine,' she said.
'I did not know he had done so. I am very much surprised to hear it.'

'Why is it surprising?'

'I thought my husband had very little money in the house. On
Sunday night, just before he went out in the motor, he came into the
drawing-room where I was sitting. He seemed to be irritated about
something, and asked me at once if I had any notes or gold I could let

him have until next day. I was surprised at that, because he was never without money; he made it a rule to carry a hundred pounds or so about him always in a note-case. I unlocked my escritoire, and gave him all I had by me. It was nearly thirty pounds.'

'And he did not tell you why he wanted it?'

'No. He put it in his pocket, and then said that Mr Marlowe had persuaded him to go for a run in the motor by moonlight, and he thought it might help him to sleep. He had been sleeping badly, as perhaps you know. Then he went off with Mr Marlowe. I thought it odd he should need money on Sunday night, but I soon forgot about it. I never remembered it again until now.'

'It was curious, certainly,' said Trent, staring into the distance. Mr Cupples began to speak to his niece of the arrangements for the inquest, and Trent moved away to where Marlowe was pacing slowly upon the lawn. The young man seemed relieved to talk about the coming business of the day. Though he still seemed tired out and nervous, he showed himself not without a quiet humour in describing the pomposities of the local police and the portentous airs of Dr Stock. Trent turned the conversation gradually toward the problem of the crime, and all Marlowe's gravity returned.

'Bunner has told me what he thinks,' he said when Trent referred to the American's theory. 'I don't find myself convinced by it, because it doesn't really explain some of the oddest facts. But I have lived long enough in the United States to know that such a stroke of revenge, done in a secret, melodramatic way, is not an unlikely thing. It is quite a characteristic feature of certain sections of the labour movement there. Americans have a taste and a talent for that sort of business. Do you know *Huckleberry Finn?*'

'Do I know my own name?' exclaimed Trent.

'Well, I think the most American thing in that great American epic is Tom Sawyer's elaboration of an extremely difficult and romantic scheme, taking days to carry out, for securing the escape of the nigger Jim, which could have been managed quite easily in twenty minutes. You know how fond they are of lodges and brotherhoods. Every college club has its secret signs and handgrips. You've heard of the Know-Nothing movement in politics, I dare say, and the Ku Klux Klan. Then look at Brigham Young's penny-dreadful tyranny in Utah, with real blood. The founders of the Mormon State were of the purest Yankee stock in America; and you know what they did. It's all part of

the same mental tendency. Americans make fun of it among themselves. For my part, I take it very seriously.'

'It can have a very hideous side to it, certainly,' said Trent, 'when
you get it in connection with crime—or with vice—or even mere
luxury. But I have a sort of sneaking respect for the determination
to make life interesting and lively in spite of civilization. To return to the matter in hand, however; has it struck you as a possibility
that Manderson's mind was affected to some extent by this
menace that Bunner believes in? For instance, it was rather an extraordinary thing to send you posting off like that in the middle of the
night.'

'About ten o'clock, to be exact,' replied Marlowe. 'Though, mind
you, if he'd actually roused me out of my bed at midnight I shouldn't
have been very much surprised. It all chimes in with what we've just
been saying. Manderson had a strong streak of the national taste for
dramatic proceedings. He was rather fond of his well-earned reputation for unexpected strokes and for going for his object with ruthless directness through every opposing consideration. He had
decided suddenly that he wanted to have word from this man
Harris—'

'Who is Harris?' interjected Trent.

'Nobody knows. Even Bunner never heard of him, and can't
imagine what the business in hand was. All I know is that when I
went up to London last week to attend to various things I booked a
deck-cabin, at Manderson's request, for a Mr George Harris on the
boat that sailed on Monday. It seems that Manderson suddenly found
he wanted news from Harris which presumably was of a character too
secret for the telegraph; and there was no train that served; so I was
sent off as you know.'

Trent looked round to make sure that they were not overheard,
then faced the other gravely, 'There is one thing I may tell you,' he
said quietly, 'that I don't think you know. Martin the butler caught a
few words at the end of your conversation with Manderson in the
orchard before you started with him in the car. He heard him say, "If
Harris is there, every moment is of importance." Now, Mr Marlowe,
you know my business here. I am sent to make enquiries, and you
mustn't take offence. I want to ask you if, in the face of that sentence,
you will repeat that you know nothing of what the business was.'

Marlowe shook his head. 'I know nothing, indeed. I'm not easily
offended, and your question is quite fair. What passed during that

conversation I have already told the detective. Manderson plainly said to me that he could not tell me what it was all about. He simply wanted me to find Harris, tell him that he desired to know how matters stood, and bring back a letter or message from him. Harris, I was further told, might not turn up. If he did, "every moment was of importance". And now you know as much as I do.'

'That talk took place *before* he told his wife that you were taking him for a moonlight run. Why did he conceal your errand in that way, I wonder.'

The young man made a gesture of helplessness. 'Why? I can guess no better than you.'

'Why,' muttered Trent as if to himself, gazing on the ground, 'did he conceal it—from Mrs Manderson?' He looked up at Marlowe.

'And from Martin,' the other amended coolly. 'He was told the same thing.'

With a sudden movement of his head Trent seemed to dismiss the subject. He drew from his breast-pocket a letter-case, and thence extracted two small leaves of clean, fresh paper.

'Just look at these two slips, Mr Marlowe,' he said. 'Did you ever see them before? Have you any idea where they come from?' he added as Marlowe took one in each hand and examined them curiously.

'They seem to have been cut with a knife or scissors from a small diary for this year—from the October pages,' Marlowe observed, looking them over on both sides. 'I see no writing of any kind on them. Nobody here has any such diary so far as I know. What about them?'

'There may be nothing in it,' Trent said dubiously. 'Any one in the house, of course, might have such a diary without your having seen it. But I didn't much expect you would be able to identify the leaves— in fact, I should have been surprised if you had.'

He stopped speaking as Mrs Manderson came towards them. 'My uncle thinks we should be going now,' she said.

'I think I will walk on with Mr Bunner,' Mr Cupples said as he joined them. 'There are certain business matters that must be disposed of as soon as possible. Will you come on with these two gentlemen, Mabel? We will wait for you before we reach the place.'

Trent turned to her. 'Mrs Manderson will excuse me, I hope,' he said. 'I really came up this morning in order to look about me here for some indications I thought I might possibly find. I had not thought of attending the—the court just yet.'

She looked at him with eyes of perfect candour. 'Of course, Mr Trent. Please do exactly as you wish. We are all relying upon you. If you will wait a few moments, Mr Marlowe, I shall be ready.'

She entered the house. Her uncle and the American had already strolled towards the gate.

Trent looked into the eyes of his companion. 'That is a wonderful woman,' he said in a lowered voice.

'You say so without knowing her,' replied Marlowe in a similar tone. 'She is more than that.'

Trent said nothing to this. He stared out over the fields towards the sea. In the silence a noise of hobnailed haste rose on the still air. A little distance down the road a boy appeared trotting towards them from the direction of the hotel. In his hand was the orange envelope, unmistakable afar off, of a telegram. Trent watched him with an indifferent eye as he met and passed the two others. Then he turned to Marlowe. 'A propos of nothing in particular,' he said, 'were you at Oxford?'

'Yes,' said the young man. 'Why do you ask?'

'I just wondered if I was right in my guess. It's one of the things you can very often tell about a man, isn't it?'

'I suppose so,' Marlowe said. 'Well, each of us is marked in one way or another, perhaps. I should have said you were an artist, if I hadn't known it.'

'Why? Does my hair want cutting?'

'Oh, no! It's only that you look at things and people as I've seen artists do, with an eye that moves steadily from detail to detail—rather looking them over than looking at them.'

The boy came up panting. 'Telegram for you, sir,' he said to Trent. 'Just come, sir.'

Trent tore open the envelope with an apology, and his eyes lighted up so visibly as he read the slip that Marlowe's tired face softened in a smile.

'It must be good news,' he murmured half to himself.

Trent turned on him a glance in which nothing could be read. 'Not exactly news,' he said. 'It only tells me that another little guess of mine was a good one.'

CHAPTER VIII

The Inquest

THE coroner, who fully realized that for that one day of his life as a provincial solicitor he was living in the gaze of the world, had resolved to be worthy of the fleeting eminence. He was a large man of jovial temper, with a strong interest in the dramatic aspects of his work, and the news of Manderson's mysterious death within his jurisdiction had made him the happiest coroner in England. A respectable capacity for marshalling facts was fortified in him by a copiousness of impressive language that made juries as clay in his hands, and sometimes disguised a doubtful interpretation of the rules of evidence.

The court was held in a long, unfurnished room lately built on to the hotel, and intended to serve as a ballroom or concert-hall. A regiment of reporters was entrenched in the front seats, and those who were to be called on to give evidence occupied chairs to one side of the table behind which the coroner sat, while the jury, in double row, with plastered hair and a spurious ease of manner, flanked him on the other side. An undistinguished public filled the rest of the space, and listened, in an awed silence, to the opening solemnities. The newspaper men, well used to these, muttered among themselves. Those of them who knew Trent by sight assured the rest that he was not in the court.

The identity of the dead man was proved by his wife, the first witness called, from whom the coroner, after some enquiry into the health and circumstances of the deceased, proceeded to draw an account of the last occasion on which she had seen her husband alive. Mrs Manderson was taken through her evidence by the coroner with the sympathy which every man felt for that dark figure of grief. She lifted her thick veil before beginning to speak, and the extreme paleness and unbroken composure of the lady produced a singular impression. This was not an impression of hardness. Interesting femininity was the first thing to be felt in her presence. She was not even enigmatic. It was only clear that the force of a powerful character was at work to master the emotions of her situation. Once or twice as she spoke she touched her eyes with her handkerchief, but her voice was low and clear to the end.

Her husband, she said, had come up to his bedroom about his usual hour for retiring on Sunday night. His room was really a dressing-room attached to her own bedroom, communicating with it by a door which was usually kept open during the night. Both dressing-room and bedroom were entered by other doors giving on the passage. Her husband had always had a preference for the greatest simplicity in his bedroom arrangements, and liked to sleep in a small room. She had not been awake when he came up, but had been half-aroused, as usually happened, when the light was switched on in her husband's room. She had spoken to him. She had no clear recollection of what she had said, as she had been very drowsy at the time; but she had remembered that he had been out for a moonlight run in the car, and she believed she had asked whether he had had a good run, and what time it was. She had asked what the time was because she felt as if she had only been a very short time asleep, and she had expected her husband to be out very late. In answer to her question he had told her it was half-past eleven, and had gone on to say that he had changed his mind about going for a run.

'Did he say why?' the coroner asked.

'Yes,' replied the lady, 'he did explain why. I remember very well what he said, because—' she stopped with a little appearance of confusion.

'Because—' the coroner insisted gently.

'Because my husband was not as a rule communicative about his business affairs,' answered the witness, raising her chin with a faint touch of defiance. 'He did not—did not think they would interest me, and as a rule referred to them as little as possible. That was why I was rather surprised when he told me that he had sent Mr Marlowe to Southampton to bring back some important information from a man who was leaving for Paris by the next day's boat. He said that Mr Marlowe could do it quite easily if he had no accident. He said that he had started in the car, and then walked back home a mile or so, and felt all the better for it.'

'Did he say any more?'

'Nothing, as well as I remember,' the witness said. 'I was very sleepy, and I dropped off again in a few moments. I just remember my husband turning his light out, and that is all. I never saw him again alive.'

'And you heard nothing in the night?'

'No: I never woke until my maid brought my tea in the morning at seven o'clock. She closed the door leading to my husband's room, as she always did, and I supposed him to be still there. He always needed a great deal of sleep. He sometimes slept until quite late in the morning. I had breakfast in my sitting-room. It was about ten when I heard that my husband's body had been found.' The witness dropped her head and silently waited for her dismissal.

But it was not to be yet.

'Mrs Manderson.' The coroner's voice was sympathetic, but it had a hint of firmness in it now. 'The question I am going to put to you must, in these sad circumstances, be a painful one; but it is my duty to ask it. Is it the fact that your relations with your late husband had not been, for some time past, relations of mutual affection and confidence? Is it the fact that there was an estrangement between you?'

The lady drew herself up again and faced her questioner, the colour rising in her cheeks. 'If that question is necessary,' she said with cold distinctness, 'I will answer it so that there shall be no misunderstanding. During the last few months of my husband's life his attitude towards me had given me great anxiety and sorrow. He had changed towards me; he had become very reserved, and seemed mistrustful. I saw much less of him than before; he seemed to prefer to be alone. I can give no explanation at all of the change. I tried to work against it; I did all I could with justice to my own dignity, as I thought. Something was between us, I did not know what, and he never told me. My own obstinate pride prevented me from asking what it was in so many words; I only made a point of being to him exactly as I had always been, so far as he would allow me. I suppose I shall never know now what it was.' The witness, whose voice had trembled in spite of her self-control over the last few sentences, drew down her veil when she had said this, and stood erect and quiet.

One of the jury asked a question, not without obvious hesitation. 'Then was there never anything of the nature of what they call Words between you and your husband, ma'am?'

'Never.' The word was colourlessly spoken; but every one felt that a crass misunderstanding of the possibilities of conduct in the case of a person like Mrs Manderson had been visited with some severity.

Did she know, the coroner asked, of any other matter which might have been preying upon her husband's mind recently?

Mrs Manderson knew of none whatever. The coroner intimated that her ordeal was at an end, and the veiled lady made her way to the door. The general attention, which followed her for a few moments, was now eagerly directed upon Martin, whom the coroner had proceeded to call.

It was at this moment that Trent appeared at the doorway and edged his way into the great room. But he did not look at Martin. He was observing the well-balanced figure that came quickly toward him along an opening path in the crowd, and his eye was gloomy. He started, as he stood aside from the door with a slight bow, to hear Mrs Manderson address him by name in a low voice. He followed her a pace or two into the hall.

'I wanted to ask you,' she said in a voice now weak and oddly broken, 'if you would give me your arm a part of the way to the house. I could not see my uncle near the door, and I suddenly felt rather faint. . . . I shall be better in the air. . . . No, no; I cannot stay here— please, Mr Trent!' she said, as he began to make an obvious suggestion. 'I must go to the house.' Her hand tightened momentarily on his arm as if, for all her weakness, she could drag him from the place; then again she leaned heavily upon it, and with that support, and with bent head, she walked slowly from the hotel and along the oak-shaded path toward White Gables.

Trent went in silence, his thoughts whirling, dancing insanely to a chorus of 'Fool! fool!' All that he alone knew, all that he guessed and suspected of this affair, rushed through his brain in a rout; but the touch of her unnerved hand upon his arm never for an instant left his consciousness, filling him with an exaltation that enraged and bewildered him. He was still cursing himself furiously behind the mask of conventional solicitude that he turned to the lady when he had attended her to the house and seen her sink upon a couch in the morning-room. Raising her veil, she thanked him gravely and frankly, with a look of sincere gratitude in her eyes. She was much better now, she said, and a cup of tea would work a miracle upon her. She hoped she had not taken him away from anything important. She was ashamed of herself; she thought she could go through with it, but she had not expected those last questions. 'I am glad you did not hear me,' she said when he explained. 'But of course you will read it all in the reports. It shook me so to have to speak of that,' she added simply; 'and to keep from making an exhibition of myself took it out of me. And all those staring men by the door! Thank you again for helping me when I asked you. . . . I thought I might,' she ended queerly, with a little tired smile; and Trent took himself away, his hand still quivering from the cool touch of her fingers.

The testimony of the servants and of the finder of the body brought nothing new to the reporters' net. That of the police was as colourless and cryptic as is usual at the inquest stage of affairs of the kind. Greatly to the satisfaction of Mr Bunner, his evidence afforded the sensation of the day, and threw far into the background the interesting revelation of domestic difficulty made by the dead man's wife. He told the court in substance what he had already told Trent. The flying pencils did not miss a word of the young American's story, and it appeared with scarcely the omission of a sentence in every journal of importance in Great Britain and the United States.

Public opinion next day took no note of the faint suggestion of the possibility of suicide which the coroner, in his final address to the jury, had thought it right to make in connection with the lady's evidence. The weight of evidence, as the official had indeed pointed out, was against such a theory. He had referred with emphasis to the fact that no weapon had been found near the body.

'This question, of course, is all-important, gentlemen,' he had said to the jury. 'It is, in fact, the main issue before you. You have seen the body for yourselves. You have just heard the medical evidence; but I think it would be well for me to read you my notes of it in so far as they bear on this point, in order to refresh your memories. Dr Stock told you—I am going to omit all technical medical language and repeat to you merely the plain English of his testimony—that in his opinion death had taken place six or eight hours previous to the finding of the body. He said that the cause of death was a bullet wound, the bullet having entered the left eye, which was destroyed, and made its way to the base of the brain, which was quite shattered. The external appearance of the wound, he said, did not support the hypothesis of its being self-inflicted, inasmuch as there were no signs of the firearm having been pressed against the eye, or even put very close to it; at the same time it was not physically impossible that the weapon should have been discharged by the deceased with his own hand, at some small distance from the eye. Dr Stock also told us that it was impossible to say with certainty, from the state of the body, whether any struggle had taken place at the time of death; that when seen by him, at which time he understood that it had not been moved since it was found, the body was lying in a collapsed position such as might very well result from the shot alone; but that the scratches and bruises upon the wrists and the lower part of the arms had been very recently inflicted, and were, in his opinion, marks of violence.

'In connection with this same point, the remarkable evidence given by Mr Bunner cannot be regarded, I think, as without significance. It may have come as a surprise to some of you to hear that risks of the character described by this witness are, in his own country, commonly run by persons in the position of the deceased. On the other hand, it may have been within the knowledge of some of you that in the industrial world of America the discontent of labour often proceeds to lengths of which we in England happily know nothing. I have interrogated the witness somewhat fully upon this. At the same time, gentlemen, I am by no means suggesting that Mr Bunner's personal conjecture as to the cause of death can fitly be adopted by you. That is emphatically not the case. What his evidence does is to raise two questions for your consideration. First, can it be said that the deceased was to any extent in the position of a threatened man—of a man more exposed to the danger of murderous attack than an ordinary person? Second, does the recent alteration in his demeanour, as described by this witness, justify the belief that his last days were overshadowed by a great anxiety? These points may legitimately be considered by you in arriving at a conclusion upon the rest of the evidence.'

Thereupon the coroner, having indicated thus clearly his opinion that Mr Bunner had hit the right nail on the head, desired the jury to consider their verdict.

CHAPTER IX

A Hot Scent

'COME in!' called Trent.

Mr Cupples entered his sitting-room at the hotel. It was the early evening of the day on which the coroner's jury, without leaving the box, had pronounced the expected denunciation of a person or persons unknown. Trent, with a hasty glance upward, continued his intent study of what lay in a photographic dish of enamelled metal, which he moved slowly about in the light of the window. He looked very pale, and his movements were nervous.

'Sit on the sofa,' he advised. 'The chairs are a job lot bought at the sale after the suppression of the Holy Inquisition in Spain. This is a pretty good negative,' he went on, holding it up to the light with his head at the angle of discriminating judgement. 'Washed enough now, I think. Let us leave it to dry, and get rid of all this mess.'

Mr Cupples, as the other busily cleared the table of a confusion of basins, dishes, racks, boxes, and bottles, picked up first one and then another of the objects and studied them with innocent curiosity.

'That is called hypo-eliminator,' said Trent, as Mr Cupples uncorked and smelt at one of the bottles. 'Very useful when you're in a hurry with a negative. I shouldn't drink it, though, all the same. It eliminates sodium hypophosphite, but I shouldn't wonder if it would eliminate human beings too.' He found a place for the last of the litter on the crowded mantel-shelf, and came to sit before Mr Cupples on the table. 'The great thing about a hotel sitting-room is that its beauty does not distract the mind from work. It is no place for the mayfly pleasures of a mind at ease. Have you ever been in this room before, Cupples? I have, hundreds of times. It has pursued me all over England for years. I should feel lost without it if, in some fantastic, far-off hotel, they were to give me some other sitting-room. Look at this table-cover; there is the ink I spilt on it when I had this room in Halifax. I burnt that hole in the carpet when I had it in Ipswich. But I see they have mended the glass over the picture of "Silent

Sympathy", which I threw a boot at in Banbury. I do all my best work here. This afternoon, for instance, since the inquest, I have finished several excellent negatives. There is a very good dark room downstairs.'

'The inquest—that reminds me,' said Mr Cupples, who knew that this sort of talk in Trent meant the excitement of action, and was wondering what he could be about. 'I came in to thank you, my dear fellow, for looking after Mabel this morning. I had no idea she was going to feel ill after leaving the box; she seemed quite unmoved, and, really, she is a woman of such extraordinary self-command, I thought I could leave her to her own devices and hear out the evidence, which I thought it important I should do. It was a very fortunate thing she found a friend to assist her, and she is most grateful. She is quite herself again now.'

Trent, with his hands in his pockets and a slight frown on his brow, made no reply to this. 'I tell you what,' he said after a short pause, 'I was just getting to the really interesting part of the job when you came in. Come; would you like to see a little bit of high-class police work? It's the very same kind of work that old Murch ought to be doing at this moment. Perhaps he is; but I hope to glory he isn't.' He sprang off the table and disappeared into his bedroom. Presently he came out with a large drawing-board on which a number of heterogeneous objects was ranged.

'First I must introduce you to these little things,' he said, setting them out on the table. 'Here is a big ivory paper-knife; here are two leaves cut out of a diary—my own diary; here is a bottle containing dentifrice; here is a little case of polished walnut. Some of these things have to be put back where they belong in somebody's bedroom at White Gables before night. That's the sort of man I am— nothing stops me. I borrowed them this very morning when every one was down at the inquest, and I dare say some people would think it rather an odd proceeding if they knew. Now there remains one object on the board. Can you tell me, without touching it, what it is?'

'Certainly I can,' said Mr Cupples, peering at it with great interest. 'It is an ordinary glass bowl. It looks like a finger-bowl. I see nothing odd about it,' he added after some moments of close scrutiny.

'I can't see much myself,' replied Trent, 'and that is exactly where the fun comes in. Now take this little fat bottle, Cupples, and pull out the cork. Do you recognize that powder inside it? You have swallowed pounds of it in your time, I expect. They give it to babies. Grey powder is its ordinary name—mercury and chalk. It is great stuff. Now, while I hold the basin sideways over this sheet of paper, I want you to pour a little powder out of the bottle over this part of the bowl—just here. . . . Perfect! Sir Edward Henry himself could not have handled the powder better. You have done this before, Cupples, I can see. You are an old hand.'

'I really am not,' said Mr Cupples seriously, as Trent returned the fallen powder to the bottle. 'I assure you it is all a complete mystery to me. What did I do then?'

'I brush the powdered part of the bowl lightly with this camel-hair brush. Now look at it again. You saw nothing odd about it before. Do you see anything now?'

Mr Cupples peered again. 'How curious!' he said. 'Yes, there are two large grey finger-marks on the bowl. They were not there before.'

'I am Hawkshaw the detective,' observed Trent. 'Would it interest you to hear a short lecture on the subject of glass finger-bowls? When you take one up with your hand you leave traces upon it, usually practically invisible, which may remain for days or months. You leave the marks of your fingers. The human hand, even when quite clean, is never quite dry, and sometimes—in moments of great anxiety, for instance, Cupples—it is very moist. It leaves a mark on any cold smooth surface it may touch. That bowl was moved by somebody with a rather moist hand quite lately.' He sprinkled the powder again. 'Here on the other side, you see, is the thumb-mark—very good impressions all of them.' He spoke without raising his voice, but Mr Cupples could perceive that he was ablaze with excitement as he stared at the faint grey marks. 'This one should be the index finger. I need not tell a man of your knowledge of the world that the pattern of it is a single-spiral whorl, with deltas symmetrically disposed. This, the print of the second finger, is a simple loop, with a staple core and fifteen counts. I know there are fifteen, because I have just the same two prints on this negative, which I have examined in detail. Look!'—he held one of the negatives up to the light of the declining sun and demonstrated with a pencil point. 'You can see they're the same. You see the bifurcation of that ridge. There it is in the other. You see that little scar near the centre. There it is in the other. There are a score of ridge-characteristics on which an expert would swear in the witness-box that the marks on that bowl and the marks I have photographed on this negative were made by the same hand.'

'And where did you photograph them? What does it all mean?' asked Mr Cupples, wide-eyed.

'I found them on the inside of the left-hand leaf of the front window in Mrs Manderson's bedroom. As I could not bring the window with me, I photographed them, sticking a bit of black paper on the other side of the glass for the purpose. The bowl comes from Manderson's room. It is the bowl in which his false teeth were placed at night. I could bring that away, so I did.'

'But those cannot be Mabel's finger-marks.'

'I should think not!' said Trent with decision. 'They are twice the size of any print Mrs Manderson could make.'

'Then they must be her husband's.'

'Perhaps they are. Now shall we see if we can match them once more? I believe we can.' Whistling faintly, and very white in the face, Trent opened another small squat bottle containing a dense black powder. 'Lamp-black,' he explained. 'Hold a bit of paper in your hand for a second or two, and this little chap will show you the pattern of your fingers.' He carefully took up with a pair of tweezers one of the leaves cut from his diary, and held it out for the other to examine. No marks appeared on the leaf. He tilted some of the powder out upon one surface of the paper, then, turning it over, upon the other; then shook the leaf gently to rid it of the loose powder. He held it out to Mr Cupples in silence. On one side of the paper appeared unmistakably, clearly printed in black, the same two finger-prints that he had already seen on the bowl and on the photographic plate. He took up the bowl and compared them. Trent turned the paper over, and on the other side was a bold black replica of the thumb-mark that was printed in grey on the glass in his hand.

'Same man, you see,' Trent said with a short laugh. 'I felt that it must be so, and now I know.' He walked to the window and looked out. 'Now I know,' he repeated in a low voice, as if to himself. His tone was bitter. Mr Cupples, understanding nothing, stared at his motionless back for a few moments.

'I am still completely in the dark,' he ventured presently. 'I have often heard of this fingerprint business, and wondered how the police went to work about it. It is of extraordinary interest to me, but upon my life I cannot see how in this case Manderson's fingerprints are going—'

'I am very sorry, Cupples,' Trent broke in upon his meditative speech with a swift return to the table. 'When I began this investigation I meant to take you with me every step of the way. You mustn't think I have any doubts about your discretion if I say now that I must hold my tongue about the whole thing, at least for a time. I will tell you this: I have come upon a fact that looks too much like having very painful consequences if it is discovered by any one else.' He looked at the other with a hard and darkened face, and struck the table with his hand. 'It is terrible for me here and now. Up to this moment I was hoping against hope that I was wrong about the fact. I may still be wrong in the surmise that I base upon that fact. There is only one way of finding out that is open to me, and I must nerve myself to take it.' He smiled suddenly at Mr Cupples's face of consternation. 'All right—I'm not going to be tragic any more, and I'll tell you all about it when I can. Look here, I'm not half through my game with the powder-bottles yet.'

He drew one of the defamed chairs to the table and sat down to test the broad ivory blade of the paper knife. Mr Cupples, swallowing his amazement, bent forward in an attitude of deep interest and handed Trent the bottle of lamp-black.

CHAPTER X

The Wife of Dives

MRS MANDERSON stood at the window of her sitting-room at White Gables gazing out upon a wavering landscape of fine rain and mist. The weather had broken as it seldom does in that part in June. White wreathings drifted up the fields from the sullen sea; the sky was an unbroken grey deadness shedding pin-point moisture that was now and then blown against the panes with a crepitation of despair. The lady looked out on the dim and chilling prospect with a woeful face. It was a bad day for a woman bereaved, alone, and without a purpose in life.

There was a knock, and she called 'Come in,' drawing herself up with an unconscious gesture that always came when she realized that the weariness of the world had been gaining upon her spirit. Mr Trent had called, the maid said; he apologized for coming at such an early hour, but hoped that Mrs Manderson would see him on a matter of urgent importance. Mrs Manderson would see Mr Trent. She walked to a mirror, looked into the olive face she saw reflected there, shook her head at herself with the flicker of a grimace, and turned to the door as Trent was shown in.

His appearance, she noted, was changed. He had the jaded look of the sleepless, and a new and reserved expression, in which her quick sensibilities felt something not propitious, took the place of his half smile of fixed good-humour.

'May I come to the point at once?' he said, when she had given him her hand. 'There is a train I ought to catch at Bishopsbridge at twelve o'clock, but I cannot go until I have settled this thing, which concerns you only, Mrs Manderson. I have been working half the night and thinking the rest; and I know now what I ought to do.'

'You look wretchedly tired,' she said kindly. 'Won't you sit down? This is a very restful chair. Of course it is about this terrible business and your work as correspondent. Please ask me anything you think I can properly tell you, Mr Trent. I know that you won't make it worse for me than you can help in doing your duty here. If you say you must see me about something, I know it must be because, as you say, you ought to do it.'

'Mrs Manderson,' said Trent, slowly measuring his words, 'I won't make it worse for you than I can help. But I am bound to make it bad for you—only between ourselves, I hope. As to whether you can properly tell me what I shall ask you, you will decide that; but I tell you this on my word of honour: I shall ask you only as much as will decide me whether to publish or to withhold certain grave things that I have found out about your husband's death, things not suspected by any one else, nor, I think, likely to be so. What I have discovered— what I believe that I have practically proved—will be a great shock to you in any case. But it may be worse for you than that; and if you give me reason to think it would be so, then I shall suppress this manuscript,' he laid a long envelope on the small table beside him, 'and nothing of what it has to tell shall ever be printed. It consists, I may tell you, of a short private note to my editor, followed by a long dispatch for publication in the *Record*. Now you may refuse to say anything to me. If you do refuse, my duty to my employers, as I see it, is to take this up to London with me today and leave it with my editor to be dealt with at his discretion. My view is, you understand, that I am not entitled to suppress it on the strength of a mere possibility that presents itself to my imagination. But if I gather from you—and I can gather it from no other person—that there is substance in that imaginary possibility I speak of, then I have only one thing to do as a gentleman and as one who'—he hesitated for a phrase—'wishes you well. I shall not publish that dispatch of mine. In some directions I decline to assist the police. Have you followed me so far?' he asked with a touch of anxiety in his careful coldness; for her face, but for its pallor, gave no sign as she regarded him, her hands clasped before her, and her shoulders drawn back in a pose of rigid calm. She looked precisely as she had looked at the inquest.

'I understand quite well,' said Mrs Manderson in a low voice. She drew a deep breath, and went on: 'I don't know what dreadful thing you have found out, or what the possibility that has occurred to you can be, but it was good, it was honourable of you to come to me about it. Now will you please tell me?'

'I cannot do that,' Trent replied. 'The secret is my newspaper's if it is not yours. If I find it is yours, you shall have my manuscript to read and destroy. Believe me,' he broke out with something of his old warmth, 'I detest such mystery-making from the bottom of my soul; but it is not I who have made this mystery. This is the most painful hour of my life, and you make it worse by not treating me like a hound. The first thing I ask you to tell me,' he reverted with an effort to his colourless tone, 'is this: is it true, as you stated at the inquest, that you had no idea at all of the reason why your late husband had changed his attitude toward you, and become mistrustful and reserved, during the last few months of his life?'

Mrs Manderson's dark brows lifted and her eyes flamed; she quickly rose from her chair. Trent got up at the same moment, and took his envelope from the table; his manner said that he perceived the interview to be at an end. But she held up a hand, and there was colour in her cheeks and quick breathing in her voice as she said: 'Do you know what you ask, Mr Trent? You ask me if I perjured myself.'

'I do,' he answered unmoved; and he added after a pause, 'you knew already that I had not come here to preserve the polite fictions, Mrs Manderson. The theory that no reputable person, being on oath, could withhold a part of the truth under any circumstances is a polite fiction.' He still stood as awaiting dismissal, but she was silent. She walked to the window, and he stood miserably watching the slight movement of her shoulders until it subsided. Then with face averted, looking out on the dismal weather, she spoke at last clearly.

'Mr Trent,' she said, 'you inspire confidence in people, and I feel that things which I don't want known or talked about are safe with you. And I know you must have a very serious reason for doing what you are doing, though I don't know what it is. I suppose it would be assisting justice in some way if I told you the truth about what you asked just now. To understand that truth you ought to know about what went before—I mean about my marriage. After all, a good many people could tell you as well as I can that it was not . . . a very successful union. I was only twenty. I admired his force and courage and certainty; he was the only strong man I had ever known. But it did not take me long to find out that he cared for his business more than for me, and I think I found out even sooner that I had been deceiving myself and blinding myself, promising myself impossible things and wilfully misunderstanding my own feelings, because I was dazzled by the idea of having more money to spend than an English girl ever dreams of. I have been despising myself for that for five years. My husband's feeling for me . . . well, I cannot speak of that . . . what I want to say is that along with it there had always been a belief of his that I was the sort of woman to take a great place in society, and that I should throw myself into it with enjoyment, and become a sort of personage and do him great credit—that was his idea; and the idea remained with him after other delusions had gone. I was a part of his ambition. That was his really bitter disappointment, that I failed him as a social success. I think he was too shrewd not to have known in his heart that such a man as he was, twenty years older than I, with great business responsibilities that filled every hour of his life, and caring for nothing else—he must have felt that there was a risk of great unhappiness in marrying the sort of girl I was, brought up to music and books and unpractical ideas, always enjoying myself in my own way. But he had really reckoned on me as a wife who would do the honours of his position in the world; and I found I couldn't.'

Mrs Manderson had talked herself into a more emotional mood than she had yet shown to Trent. Her words flowed freely, and her voice had begun to ring and give play to a natural expressiveness that must hitherto have been dulled, he thought, by the shock and self-restraint of the past few days. Now she turned swiftly from the window and faced him as she went on, her beautiful face flushed and animated, her eyes gleaming, her hands moving in slight emphatic gestures, as she surrendered herself to the impulse of giving speech to things long pent up.

'The people,' she said. 'Oh, those people! Can you imagine what it must be for any one who has lived in a world where there was always creative work in the background, work with some dignity about it, men and women with professions or arts to follow, with ideals and things to believe in and quarrel about, some of them wealthy, some of them quite poor; can you think what it means to step out of that into another world where you *have* to be very rich, shamefully rich, to exist at all—where money is the only thing that counts and the first thing in everybody's thoughts—where the men who make the millions are so jaded by the work, that sport is the only thing they can occupy themselves with when they have any leisure, and the men who don't have to work are even duller than the men who do, and vicious as well; and the women live for display and silly amusements and silly immoralities; do you know how awful that life is? Of course I know there are clever people, and people of taste in that set, but they're swamped and spoiled, and it's the same thing in the end; empty, empty! Oh! I suppose I'm exaggerating, and I did make friends and have some happy times; but that's how I feel after it all. The seasons in New York and London—how I hated them! And our house-parties and cruises in the yacht and the rest—the same people, the same emptiness.

'And you see, don't you, that my husband couldn't have an idea of all this. *His* life was never empty. He did not live it in society, and when he was in society he had always his business plans and difficulties to occupy his mind. He hadn't a suspicion of what I felt, and I never let him know; I couldn't, it wouldn't have been fair. I felt I must do *something* to justify myself as his wife, sharing his position and fortune; and the only thing I could do was to try, and try, to live up to his idea about my social qualities . . . I did try. I acted my best. And it became harder year by year . . . I never was what they call a popular hostess, how could I be? I was a failure; but I went on trying . . . I used to steal holidays now and then. I used to feel as if I was not doing my part of a bargain—it sounds horrid to put it like that, I know, but it *was* so—when I took one of my old school-friends, who couldn't afford to travel, away to Italy for a month or two, and we went about cheaply all by ourselves, and were quite happy; or when I went and made a long stay in London with some quiet people who had known me all my life, and we all lived just as in the old days, when we had to think twice about seats at the theatre, and told each other about cheap dressmakers. Those and a few other expeditions of the same sort were my best times after I was married, and they helped me to go through with it the rest of the time. But I felt my husband would have hated to know how much I enjoyed every hour of those returns to the old life.

'And in the end, in spite of everything I could do, he came to know. . . . He could see through anything, I think, once his attention was turned to it. He had always been able to see that I was not fulfilling his idea of me as a figure in the social world, and I suppose he thought it was my misfortune rather than my fault. But the moment he began to see, in spite of my pretending, that I wasn't playing my part with any spirit, he knew the whole story; he divined how I loathed and was weary of the luxury and the brilliancy and the masses of money just because of the people who lived among them—who were made so by them, I suppose. . . . It happened last year. I don't know just how or when. It may have been suggested to him by some woman—for *they* all understood, of course. He said nothing to me, and I think he tried not to change in his manner to me at first; but such things hurt—and it was working in both of us. I knew that he knew. After a time we were just being polite and considerate to each other. Before he found me out we had been on a footing of—how can I express it to you?—of intelligent companionship, I might say. We talked without restraint of many things of the kind we could agree or disagree about without its going very deep . . . if you understand. And then that came to an end. I felt that the only possible basis of our living in each other's company was going under my feet. And at last it was gone.

'It had been like that,' she ended simply, 'for months before he died.' She sank into the corner of a sofa by the window, as though relaxing her body after an effort. For a few moments both were silent. Trent was hastily sorting out a tangle of impressions. He was amazed at the frankness of Mrs Manderson's story. He was amazed at the vigorous expressiveness in her telling of it. In this vivid being, carried away by an impulse to speak, talking with her whole personality, he had seen the real woman in a temper of activity, as he had already seen the real woman by chance in a temper of reverie and unguarded emotion. In both she was very unlike the pale, self-disciplined creature of majesty that she had been to the world. With that amazement of his went something like terror of her dark beauty, which excitement kindled into an appearance scarcely mortal in his eyes. Incongruously there rushed into his mind, occupied as it was with the affair of the moment, a little knot of ideas . . . she was unique not because of her beauty but because of its being united with intensity of nature; in England all the very beautiful women were placid, all the fiery women seemed to have burnt up the best of their beauty; that was why no beautiful woman had ever cast this sort of spell on him before; when it was a question of wit in women he had preferred the brighter flame to the duller, without much regarding the lamp. 'All this is very disputable,' said his reason; and instinct answered, 'Yes, except that I am under a spell'; and a deeper instinct cried out, 'Away with it!' He forced his mind back to her story, and found growing swiftly in him an irrepressible conviction. It was all very fine; but it would not do.

'I feel as if I had led you into saying more than you meant to say, or than I wanted to learn,' he said slowly. 'But there is one brutal question which is the whole point of my enquiry.' He braced his frame like one preparing for a plunge into cold waters. 'Mrs Manderson, will you assure me that your husband's change toward you had nothing to do with John Marlowe?'

And what he had dreaded came. 'Oh!' she cried with a sound of anguish, her face thrown up and open hands stretched out as if for pity; and then the hands covered the burning face, and she flung herself aside among the cushions at her elbow, so that he saw nothing but her heavy crown of black hair, and her body moving with sobs that stabbed his heart, and a foot turned inward gracelessly in an abandonment of misery. Like a tall tower suddenly breaking apart she had fallen in ruins, helplessly weeping.

Trent stood up, his face white and calm. With a senseless particularity he placed his envelope exactly in the centre of the little polished table. He walked to the door, closed it noiselessly as he went out, and in a few minutes was tramping through the rain out of sight of White Gables, going nowhere, seeing nothing, his soul shaken in the fierce effort to kill and trample the raving impulse that had seized him in the presence of her shame, that clamoured to him to drag himself before her feet, to pray for pardon, to pour out words—he knew not what words, but he knew that they had been straining at his lips—to wreck his self-respect for ever, and hopelessly defeat even the crazy purpose that had almost possessed him, by drowning her wretchedness in disgust, by babbling with the tongue of infatuation to a woman with a husband not yet buried, to a woman who loved another man.

Such was the magic of her tears, quickening in a moment the thing which, as his heart had known, he must not let come to life. For Philip Trent was a young man, younger in nature even than his years, and a way of life that kept his edge keen and his spirit volcanic had prepared him very ill for the meeting that comes once in the early manhood of most of us, usually—as in his case, he told himself harshly—to no purpose but the testing of virtue and the power of the will.

CHAPTER XI

Hitherto Unpublished

MY DEAR MOLLOY:—This is in case I don't find you at your office. I have found out who killed Manderson, as this dispatch will show. This was my problem; yours is to decide what use to make of it. It definitely charges an unsuspected person with having a hand in the crime, and practically accuses him of being the murderer, so I don't suppose you will publish it before his arrest, and I believe it is illegal to do so afterwards until he has been tried and found guilty. You may decide to publish it then; and you may find it possible to make some use or other before then of the facts I have given. That is your affair. Meanwhile, will you communicate with Scotland Yard, and let them see what I have written? I have done with the Manderson mystery, and I wish to God I had never touched it. Here follows my dispatch.—P.T.

Marlstone, *June* 16*th.*

I begin this, my third and probably my final dispatch to the *Record* upon the Manderson murder, with conflicting feelings. I have a strong sense of relief, because in my two previous dispatches I was obliged, in the interests of justice, to withhold facts ascertained by me which would, if published then, have put a certain person upon his guard and possibly have led to his escape; for he is a man of no common boldness and resource. These facts I shall now set forth. But I have, I confess, no liking for the story of treachery and perverted cleverness which I have to tell. It leaves an evil taste in the mouth, a savour of something revolting in the deeper puzzle of motive underlying the puzzle of the crime itself, which I believe I have solved.

It will be remembered that in my first dispatch I described the situation as I found it on reaching this place early on Tuesday morning. I told how the body was found, and in what state; dwelt upon the complete mystery surrounding the crime, and mentioned one or two local theories about it; gave some account of the dead man's domestic surroundings; and furnished a somewhat detailed description of his movements on the evening before his death. I gave, too, a little fact which may or may not have seemed irrelevant: that a quantity of whisky much larger than Manderson habitually drank at night had disappeared from his private decanter since the last time he was seen alive. On the following day, the day of the inquest, I wired little more than an abstract of the proceedings in the coroner's court, of which a verbatim report was made at my request by other representatives of the *Record*. That day is not yet over as I write these lines; and I have now completed an investigation which has led me directly to the man who must be called upon to clear himself of the guilt of the death of Manderson.

Apart from the central mystery of Manderson's having arisen long before his usual hour to go out and meet his death, there were two minor points of oddity about this affair which, I suppose, must have occurred to thousands of those who have read the accounts in the newspapers: points apparent from the very beginning. The first of these was that, whereas the body was found at a spot not thirty yards from the house, all the people of the house declared that they had heard no cry or other noise in the night. Manderson had not been gagged; the marks on his wrists pointed to a struggle with his assailant; and there had been at least one pistol-shot. (I say at least one, because it is the fact that in murders with firearms, especially if there has been a struggle, the criminal commonly misses his victim at least once.) This odd fact seemed all the more odd to me when I learned that Martin the butler was a bad sleeper, very keen of hearing, and that his bedroom, with the window open, faced almost directly toward the shed by which the body was found.

The second odd little fact that was apparent from the outset was Manderson's leaving his dental plate by the bedside. It appeared that he had risen and dressed himself fully, down to his necktie and watch and chain, and had gone out of doors without remembering to put in this plate, which he had carried in his mouth every day for years, and which contained all the visible teeth of the upper jaw. It had evidently not been a case of frantic hurry; and even if it had been, he would have been more likely to forget almost anything than this denture. Any one who wears such a removable plate will agree that the putting it in on rising is a matter of second nature. Speaking as well as eating, to say nothing of appearances, depend upon it.

Neither of these queer details, however, seemed to lead to anything at the moment. They only awakened in me a suspicion of something lurking in the shadows, something that lent more mystery to the already mysterious question how and why and through whom Manderson met his end.

With this much of preamble I come at once to the discovery which, in the first few hours of my investigation, set me upon the path which so much ingenuity had been directed to concealing.

I have already described Manderson's bedroom, the rigorous simplicity of its furnishing, contrasted so strangely with the multitude of clothes and shoes, and the manner of its communication with Mrs Manderson's room. On the upper of the two long shelves on which the shoes were ranged I found, where I had been told I should find them, the pair of patent leather shoes which Manderson had worn on the evening before his death. I had glanced over the row, not with any idea of their giving me a clue, but merely because it happens that I am a judge of shoes, and all these shoes were of the very best workmanship. But my attention was at once caught by a little peculiarity in this particular pair. They were the lightest kind of lace-up dress shoes, very thin in the sole, without toe-caps, and beautifully made, like all the rest. These shoes were old and well worn; but being carefully polished, and fitted, as all the shoes were, upon their trees, they looked neat enough. What caught my eye was a slight splitting of the leather in that part of the upper known as the vamp—a splitting at the point where the two laced parts of the shoe rise from the upper. It is at this point that the strain comes when a tight shoe of this sort is forced upon the foot, and it is usually guarded with a strong stitching across the bottom of the opening. In both the shoes I was examining this stitching had parted, and the leather below had given way. The splitting was a tiny affair in each case, not an eighth of an inch long, and the torn edges having come together again on the removal of the strain, there was nothing that a person who was not something of a connoisseur of shoe-leather would have noticed. Even less noticeable, and indeed not to be seen at all unless one were looking for it, was a slight straining of the stitches uniting the upper to the sole. At the toe and on the outer side of each shoe this stitching had been dragged until it was visible on a close inspection of the join.

These indications, of course, could mean only one thing—the shoes had been worn by some one for whom they were too small.

Now it was clear at a glance that Manderson was always thoroughly well shod, and careful, perhaps a little vain, of his small and narrow feet. Not one of the other shoes in the collection, as I soon ascertained, bore similar marks; they had not belonged to a man who squeezed himself into tight shoe-leather. Some one who was not Manderson had worn these shoes, and worn them recently; the edges of the tears were quite fresh.

The possibility of some one having worn them since Manderson's death was not worth considering; the body had only been found about twenty-six hours when I was examining the shoes; besides, why should any one wear them? The possibility of some one having borrowed Manderson's shoes and spoiled them for him while he was alive seemed about as negligible. With others to choose from he would not have worn these. Besides, the only men in the place were the butler and the two secretaries. But I do not say that I gave those possibilities even as much consideration as they deserved, for my thoughts were running away with me, and I have always found it good policy, in cases of this sort, to let them have their heads. Ever since I had got out of the train at Marlstone early that morning I had been steeped in details of the Manderson affair; the thing had not once been out of my head. Suddenly the moment had come when the dæmon wakes and begins to range.

Let me put it less fancifully. After all, it is a detail of psychology familiar enough to all whose business or inclination brings them in contact with difficult affairs of any kind. Swiftly and spontaneously, when chance or effort puts one in possession of the key-fact in any system of baffling circumstances, one's ideas seem to rush to group themselves anew in relation to that fact, so that they are suddenly rearranged almost before one has consciously grasped the significance of the key-fact itself. In the present instance, my brain had scarcely formulated within itself the thought, 'Somebody who was not Manderson has been wearing these shoes,' when there flew into my mind a flock of ideas, all of the same character and all bearing upon this new notion. It was unheard-of for Manderson to drink much whisky at night. It was very unlike him to be untidily dressed, as the body was when found—the cuffs dragged up inside the sleeves, the shoes unevenly laced; very unlike him not to wash when he rose, and to put on last night's evening shirt and collar and underclothing; very unlike him to have his watch in the waistcoat pocket that was not lined with leather for its reception. (In my first dispatch I mentioned all these points, but neither I nor any one else saw anything significant in them when examining the body.) It was very strange, in the existing domestic situation, that Manderson should be communicative to his wife about his doings, especially at the time of his going to bed, when he seldom spoke to her at all. It was extraordinary that Manderson should leave his bedroom without his false teeth.

All these thoughts, as I say, came flocking into my mind together, drawn from various parts of my memory of the morning's enquiries and observations. They had all presented themselves, in far less time than it takes to read them as set down here, as I was turning over the shoes, confirming my own certainty on the main point. And yet when I confronted the definite idea that had sprung up suddenly and unsupported before me—'*It was not Manderson who was in the house that night*'—it seemed a stark absurdity at the first formulating. It was certainly Manderson who had dined at the house and gone out with Marlowe in the car. People had seen him at close quarters. But was it he who returned at ten? That question too seemed absurd enough. But I could not set it aside. It seemed to me as if a faint light was beginning to creep over the whole expanse of my mind, as it does over land at dawn, and that presently the sun would be rising. I set myself to think over, one by one, the points that had just occurred to me, so as to make out, if possible, why any man masquerading as Manderson should have done these things that Manderson would not have done.

I had not to cast about very long for the motive a man might have in forcing his feet into Manderson's narrow shoes. The examination of footmarks is very well understood by the police. But not only was the man concerned to leave no footmarks of his own: he was concerned to leave Manderson's, if any; his whole plan, if my guess was right, must have been directed to producing the belief that Manderson was in the place that night. Moreover, his plan did not turn upon leaving footmarks. He meant to leave the shoes themselves, and he did so. The maidservant had found them outside the bedroom door, as Manderson always left his shoes, and had polished them, replacing them on the shoe-shelves later in the morning, after the body had been found.

When I came to consider in this new light the leaving of the false teeth, an explanation of what had seemed the maddest part of the affair broke upon me at once. A dental plate is not inseparable from its owner. If my guess was right, the unknown had brought the denture to the house with him, and left it in the bedroom, with the same object as he had in leaving the shoes: to make it impossible that any one should doubt that Manderson had been in the house and had gone to bed there. This, of course, led me to the inference that *Manderson was dead before the false Manderson came to the house*; and other things confirmed this.

For instance, the clothing, to which I now turned in my review of the position. If my guess was right, the unknown in Manderson's shoes had certainly had possession of Manderson's trousers, waist-coat, and shooting jacket. They were there before my eyes in the bedroom; and Martin had seen the jacket—which nobody could have mistaken—upon the man who sat at the telephone in the library. It was now quite plain (if my guess was right) that this unmistakable garment was a cardinal feature of the unknown's plan. He knew that Martin would take him for Manderson at the first glance.

And there my thinking was interrupted by the realization of a thing that had escaped me before. So strong had been the influence of the unquestioned assumption that it was Manderson who was present that night, that neither I nor, as far as I know, any one else had noted the point. *Martin had not seen the man's face, nor had Mrs Manderson.*

Mrs Manderson (judging by her evidence at the inquest, of which, as I have said, I had a full report made by the *Record* stenographers in court) had not seen the man at all. She hardly could have done, as I shall show presently. She had merely spoken with him as she lay half asleep, resuming a conversation which she had had with her living husband about an hour before. Martin, I perceived, could only have seen the man's back, as he sat crouching over the telephone; no doubt a characteristic pose was imitated there. And the man had worn his hat, Manderson's broad-brimmed hat! There is too much character in the back of a head and neck. The unknown, in fact, supposing him to have been of about Manderson's build, had had no need for any disguise, apart from the jacket and the hat and his powers of mimicry.

I paused there to contemplate the coolness and ingenuity of the man. The thing, I now began to see, was so safe and easy, provided that his mimicry was good enough, and that his nerve held. Those two points assured, only some wholly unlikely accident could unmask him.

To come back to my puzzling out of the matter as I sat in the dead man's bedroom with the tell-tale shoes before me. The reason for the entrance by the window instead of by the front door will already have occurred to any one reading this. Entering by the door, the man would almost certainly have been heard by the sharp-eared Martin in his pantry just across the hall; he might have met him face to face.

Then there was the problem of the whisky. I had not attached much importance to it; whisky will sometimes vanish in very queer ways in a household of eight or nine persons; but it had seemed strange that it should go in that way on that evening. Martin had been plainly quite dumbfounded by the fact. It seemed to me now that many a man—fresh, as this man in all likelihood was, from a bloody business, from the unclothing of a corpse, and with a desperate part still to play—would turn to that decanter as to a friend. No doubt he had a drink before sending for Martin; after making that trick with ease and success, he probably drank more.

But he had known when to stop. The worst part of the enterprise was before him: the business—clearly of such vital importance to him, for whatever reason—of shutting himself in Manderson's room and preparing a body of convincing evidence of its having been occupied by Manderson; and this with the risk—very slight, as no doubt he understood, but how unnerving!—of the woman on the other side of the half-open door awaking and somehow discovering him. True, if he kept out of her limited field of vision from the bed, she could only see him by getting up and going to the door. I found that to a person lying in her bed, which stood with its head to the wall a little beyond the door, nothing was visible through the doorway but one of the cupboards by Manderson's bed-head. Moreover, since this man knew the ways of the household, he would think it most likely that Mrs Manderson was asleep. Another point with him, I guessed, might have been the estrangement between the husband and wife, which they had tried to cloak by keeping up, among other things, their usual practice of sleeping in connected rooms, but which was well known to all who had anything to do with them. He would hope from this that if Mrs Manderson heard him, she would take no notice of the supposed presence of her husband.

So, pursuing my hypothesis, I followed the unknown up to the bedroom, and saw him setting about his work. And it was with a catch in my own breath that I thought of the hideous shock with which he must have heard the sound of all others he was dreading most: the drowsy voice from the adjoining room.

What Mrs Manderson actually said, she was unable to recollect at the inquest. She thinks she asked her supposed husband whether he had had a good run in the car. And now what does the unknown do? Here, I think, we come to a supremely significant point. Not only does he—standing rigid there, as I picture him, before the dressing-table, listening to the sound of his own leaping heart—not only does he answer the lady in the voice of Manderson; he volunteers an explanatory statement. He tells her that he has, on a sudden inspira-tion, sent Marlowe in the car to Southampton; that he has sent him to bring back some important information from a man leaving for Paris by the steamboat that morning. Why these details from a man who had long been uncommunicative to his wife, and that upon a point scarcely likely to interest her? Why these details *about Marlowe*?

Having taken my story so far, I now put forward the following definite propositions: that between a time somewhere about ten, when the car started, and a time somewhere about eleven, Manderson was shot—probably at a considerable distance from the house, as no shot was heard; that the body was brought back, left by the shed, and stripped of its outer clothing; that at some time round about eleven o'clock a man who was not Manderson, wearing Manderson's shoes, hat, and jacket, entered the library by the garden window; that he had with him Manderson's black trousers, waistcoat, and motor-coat, the denture taken from Manderson's mouth, and the weapon with which he had been murdered; that he concealed these, rang the bell for the butler, and sat down at the telephone with his hat on and his back to the door; that he was occupied with the telephone all the time Martin was in the room; that on going up to the bedroom floor he quietly entered Marlowe's room and placed the revolver with which the crime had been committed—Marlowe's revolver—in the case on the mantelpiece from which it had been taken; and that he then went to Manderson's room, placed Manderson's shoes outside the door, threw Manderson's garments on a chair, placed the denture in the bowl by the bedside, and selected a suit of clothes, a pair of shoes, and a tie from those in the bedroom.

Here I will pause in my statement of this man's proceedings to go into a question for which the way is now sufficiently prepared:

Who was the false Manderson?

Reviewing what was known to me, or might almost with certainty be surmised, about that person, I set down the following five conclusions:

(1.) He had been in close relations with the dead man. In his acting before Martin and his speaking to Mrs Manderson he had made no mistake.

(2.) He was of a build not unlike Manderson's, especially as to height and breadth of shoulder, which mainly determine the character of the back of a seated figure when the head is concealed and the body loosely clothed. But his feet were larger, though not greatly larger, than Manderson's.

(3.) He had considerable aptitude for mimicry and acting—probably some experience too.

(4.) He had a minute acquaintance with the ways of the Manderson household.

(5.) He was under a vital necessity of creating the belief that Manderson was alive and in that house until some time after midnight on the Sunday night.

So much I took as either certain or next door to it. It was as far as I could see. And it was far enough.

I proceed to give, in an order corresponding with the numbered paragraphs above, such relevant facts as I was able to obtain about Mr John Marlowe, from himself and other sources:

(1.) He had been Mr Manderson's private secretary, upon a footing of great intimacy, for nearly four years.

(2.) The two men were nearly of the same height, about five feet eleven inches; both were powerfully built and heavy in the shoulder. Marlowe, who was the younger by some twenty years, was rather slighter about the body, though Manderson was a man in good physical condition. Marlowe's shoes (of which I examined several pairs) were roughly about one shoemaker's size longer and broader than Manderson's.

(3.) In the afternoon of the first day of my investigation, after arriving at the results already detailed, I sent a telegram to a personal friend, a Fellow of a college at Oxford, whom I knew to be interested in theatrical matters, in these terms:

Please wire John Marlowe's record in connection with acting at Oxford some time past decade very urgent and confidential.

My friend replied in the following telegram, which reached me next morning (the morning of the inquest):

Marlowe was member O.U.D.S for three years and president 19– played Bardolph Cleon and Mercutio excelled in character acting and imitations in great demand at smokers was hero of some historic hoaxes.

I had been led to send the telegram which brought this very helpful answer by seeing on the mantel-shelf in Marlowe's bedroom a photograph of himself and two others in the costume of Falstaff's three followers, with an inscription from *The Merry Wives*, and by noting that it bore the imprint of an Oxford firm of photographers.

(4.) During his connection with Manderson, Marlowe had lived as one of the family. No other person, apart from the servants, had his opportunities for knowing the domestic life of the Mandersons in detail.

(5.) I ascertained beyond doubt that Marlowe arrived at a hotel in Southampton on the Monday morning at 6.30, and there proceeded to carry out the commission which, according to his story, and according to the statement made to Mrs Manderson in the bedroom by the false Manderson, had been entrusted to him by his employer. He had then returned in the car to Marlstone, where he had shown great amazement and horror at the news of the murder.

These, I say, are the relevant facts about Marlowe. We must now examine fact number 5 (as set out above) in connection with conclusion number 5 about the false Manderson.

I would first draw attention to one important fact. *The only person who professed to have heard Manderson mention Southampton at all before he started in the car was Marlowe.* His story—confirmed to some extent by what the butler overheard—was that the journey was all arranged in a private talk before they set out, and he could not say, when I put the question to him, why Manderson should have concealed his intentions by giving out that he was going with Marlowe for a moonlight drive. This point, however, attracted no attention. Marlowe had an absolutely air-tight alibi in his presence at Southampton by 6.30; nobody thought of him in connection with a murder which must have been committed after 12.30—the hour at which Martin the butler had gone to bed. But it was the Manderson who came back from the drive who went out of his way to mention Southampton openly to two persons. *He even went so far as to ring up a hotel at Southampton and ask questions which bore out Marlowe's story of his errand.* This was the call he was busy with when Martin was in the library.

Now let us consider the alibi. If Manderson was in the house that night, and if he did not leave it until some time after 12.30, Marlowe could not by any possibility have had a direct hand in the murder. It is a question of the distance between Marlstone and Southampton. If he had left Marlstone in the car at the hour when he is supposed to have done so—between 10 and 10.30—with a message from Manderson, the run would be quite an easy one to do in the time. But it would be physically impossible for the car—a 15 h.p. four-cylinder Northumberland, an average medium-power car—to get to South-ampton by half-past six unless it left Marlstone by midnight at latest. Motorists who will examine the road-map and make the calculations required, as I did in Manderson's library that day, will agree that on the facts as they appeared there was absolutely no case against Marlowe.

But even if they were not as they appeared; if Manderson was dead by eleven o'clock, and if at about that time Marlowe impersonated him at White Gables; if Marlowe retired to Manderson's bedroom—how can all this be reconciled with his appearance next morning at Southampton? *He had to get out of the house, unseen and unheard, and away in the car by midnight.* And Martin, the sharp-eared Martin, was sitting up until 12.30 in his pantry, with the door open, listening for the telephone bell. Practically he was standing sentry over the foot of the staircase, the only staircase leading down from the bedroom floor.

With this difficulty we arrive at the last and crucial phase of my investigation. Having the foregoing points clearly in mind, I spent the rest of the day before the inquest in talking to various persons and in going over my story, testing it link by link. I could only find the one weakness which seemed to be involved in Martin's sitting up until 12.30; and since his having been instructed to do so was certainly a part of the plan, meant to clinch the alibi for Marlowe, I knew there must be an explanation somewhere. If I could not find that explanation, my theory was valueless. I must be able to show that at the time Martin went up to bed the man who had shut himself in Manderson's bedroom might have been many miles away on the road to Southampton.

I had, however, a pretty good idea already—as perhaps the reader of these lines has by this time, if I have made myself clear—of how the escape of the false Manderson before midnight had been con-

trived. But I did not want what I was now about to do to be known.
If I had chanced to be discovered at work, there would have been no
concealing the direction of my suspicions. I resolved not to test them
on this point until the next day, during the opening proceedings at
the inquest. This was to be held, I knew, at the hotel, and I reckoned
upon having White Gables to myself so far as the principal inmates
were concerned.

So in fact it happened. By the time the proceedings at the hotel had
begun I was hard at work at White Gables. I had a camera with me. I
made search, on principles well known to and commonly practised by
the police, and often enough by myself, for certain indications. Without describing my search, I may say at once that I found and was able
to photograph two fresh fingerprints, very large and distinct, on the
polished front of the right-hand top drawer of the chest of drawers in
Manderson's bedroom; five more (among a number of smaller and
less recent impressions made by other hands) on the glasses of the
French window in Mrs Manderson's room, a window which always
stood open at night with a curtain before it; and three more upon
the glass bowl in which Manderson's dental plate had been found
lying.

I took the bowl with me from White Gables. I took also a few
articles which I selected from Marlowe's bedroom, as bearing the
most distinct of the innumerable fingerprints which are always to be
found upon toilet articles in daily use. I already had in my possession,
made upon leaves cut from my pocket diary, some excellent
fingerprints of Marlowe's which he had made in my presence without
knowing it. I had shown him the leaves, asking if he recognized them;
and the few seconds during which he had held them in his fingers had
sufficed to leave impressions which I was afterwards able to bring out.

By six o'clock in the evening, two hours after the jury had brought
in their verdict against a person or persons unknown, I had completed
my work, and was in a position to state that two of the five large prints
made on the window-glasses, and the three on the bowl, were made
by the left hand of Marlowe; that the remaining three on the window
and the two on the drawer were made by his right hand.

By eight o'clock I had made at the establishment of Mr H. T.
Copper, photographer, of Bishopsbridge, and with his assistance, a
dozen enlarged prints of the finger-marks of Marlowe, clearly showing the identity of those which he unknowingly made in my presence

and those left upon articles in his bedroom, with those found by me as I have described, and thus establishing the facts that Marlowe was recently in Manderson's bedroom, where he had in the ordinary way no business, and in Mrs Manderson's room, where he had still less. I hope it may be possible to reproduce these prints for publication with this dispatch.

At nine o'clock I was back in my room at the hotel and sitting down to begin this manuscript. I had my story complete.

I bring it to a close by advancing these further propositions: that on the night of the murder the impersonator of Manderson, being in Manderson's bedroom, told Mrs Manderson, as he had already told Martin, that Marlowe was at that moment on his way to Southampton; that having made his dispositions in the room, he switched off the light, and lay in the bed in his clothes; that he waited until he was assured that Mrs Manderson was asleep; that he then arose and stealthily crossed Mrs Manderson's bedroom in his stocking feet, having under his arm the bundle of clothing and shoes for the body; that he stepped behind the curtain, pushing the doors of the window a little further open with his hands, strode over the iron railing of the balcony, and let himself down until only a drop of a few feet separated him from the soft turf of the lawn.

All this might very well have been accomplished within half an hour of his entering Manderson's bedroom, which, according to Martin, he did at about half-past eleven.

What followed your readers and the authorities may conjecture for themselves. The corpse was found next morning clothed—rather untidily. Marlowe in the car appeared at Southampton by half-past six.

I bring this manuscript to an end in my sitting-room at the hotel at Marlstone. It is four o'clock in the morning. I leave for London by the noon train from Bishopsbridge, and immediately after arriving I shall place these pages in your hands. I ask you to communicate the substance of them to the Criminal Investigation Department.

PHILIP TRENT.

CHAPTER XII

Evil Days

'I AM returning the cheque you sent for what I did on the Manderson case,' Trent wrote to Sir James Molloy from Munich, whither he had gone immediately after handing in at the *Record* office a brief dispatch bringing his work on the case to an unexciting close. 'What I sent you wasn't worth one-tenth of the amount; but I should have no scruple about pocketing it if I hadn't taken a fancy—never mind why—not to touch any money at all for this business. I should like you, if there is no objection, to pay for the stuff at your ordinary space-rate, and hand the money to some charity which does not devote itself to bullying people, if you know of any such. I have come to this place to see some old friends and arrange my ideas, and the idea that comes out uppermost is that for a little while I want some employment with activity in it. I find I can't paint at all: I couldn't paint a fence. Will you try me as your Own Correspondent somewhere? If you can find me a good adventure I will send you good accounts. After that I could settle down and work.'

Sir James sent him instructions by telegram to proceed at once to Kurland and Livonia, where Citizen Browning was abroad again, and town and countryside blazed in revolt. It was a roving commission, and for two months Trent followed his luck. It served him not less well than usual. He was the only correspondent who saw General Dragilew killed in the street at Volmar by a girl of eighteen. He saw burnings, lynchings, fusillades, hangings; each day his soul sickened afresh at the imbecilities born of misrule. Many nights he lay down in danger. Many days he went fasting. But there was never an evening or a morning when he did not see the face of the woman whom he hopelessly loved.

He discovered in himself an unhappy pride at the lasting force of this infatuation. It interested him as a phenomenon; it amazed and enlightened him. Such a thing had not visited him before. It confirmed so much that he had found dubious in the recorded experience of men.

It was not that, at thirty-two, he could pretend to ignorance of this world of emotion. About his knowledge let it be enough to say that what he had learned had come unpursued and unpurchased, and was without intolerable memories; broken to the realities of sex, he was still troubled by its inscrutable history. He went through life full of a strange respect for certain feminine weakness and a very simple terror of certain feminine strength. He had held to a rather lukewarm faith that something remained in him to be called forth, and that the voice that should call would be heard in its own time, if ever, and not through any seeking.

But he had not thought of the possibility that, if this proved true some day, the truth might come in a sinister shape. The two things that had taken him utterly by surprise in the matter of his feeling towards Mabel Manderson were the insane suddenness of its uprising in full strength and its extravagant hopelessness. Before it came, he had been much disposed to laugh at the permanence of unrequited passion as a generous boyish delusion. He knew now that he had been wrong, and he was living bitterly in the knowledge.

Before the eye of his fancy the woman always came just as she was when he had first had sight of her, with the gesture which he had surprised as he walked past unseen on the edge of the cliff; that great gesture of passionate joy in her new liberty which had told him more plainly than speech that her widowhood was a release from torment, and had confirmed with terrible force the suspicion, active in his mind before, that it was her passport to happiness with a man whom she loved. He could not with certainty name to himself the moment when he had first suspected that it might be so. The seed of the thought must have been sown, he believed, at his first meeting with Marlowe; his mind would have noted automatically that such evident strength and grace, with the sort of looks and manners that the tall young man possessed, might go far with any woman of unfixed affections. And the connection of this with what Mr Cupples had told him of the Mandersons' married life must have formed itself in the unconscious depths of his mind. Certainly it had presented itself as an already established thing when he began, after satisfying himself of the identity of the murderer, to cast about for the motive of the crime. Motive, motive! How desperately he had sought for another, turning his back upon that grim thought, that Marlowe—obsessed by passion like himself, and privy perhaps to maddening truths about the wife's unhappiness—had taken a leaf, the guiltiest, from the book of Bothwell. But in all his investigations at the time, in all his broodings on the matter afterwards, he had been able to discover nothing that could prompt Marlowe to such a deed—nothing but that temptation, the whole strength of which he could not know, but which if it had existed must have pressed urgently upon a bold spirit in which scruple had been somehow paralysed. If he could trust his senses at all, the young man was neither insane nor by nature evil. But that could not clear him. Murder for a woman's sake, he thought, was not a rare crime, Heaven knew! If the modern feebleness of impulse in the comfortable classes, and their respect for the modern apparatus of detection, had made it rare among them, it was yet far from impossible. It only needed a man of equal daring and intelligence, his soul drugged with the vapours of an intoxicating intrigue, to plan and perform such a deed.

A thousand times, with a heart full of anguish, he had sought to reason away the dread that Mabel Manderson had known too much of what had been intended against her husband's life. That she knew all the truth after the thing was done he could not doubt; her unforgettable collapse in his presence when the question about Marlowe was suddenly and bluntly put, had swept away his last hope that there was no love between the pair, and had seemed to him, moreover, to speak of dread of discovery. In any case, she knew the truth after reading what he had left with her; and it was certain that no public suspicion had been cast upon Marlowe since. She had destroyed his manuscript, then, and taken him at his word to keep the secret that threatened her lover's life.

But it was the monstrous thought that she might have known murder was brewing, and guiltily kept silence, that haunted Trent's mind. She might have suspected, have guessed something; was it conceivable that she was aware of the whole plot, that she connived? He could never forget that his first suspicion of Marlowe's motive in the crime had been roused by the fact that his escape was made through the lady's room. At that time, when he had not yet seen her, he had been ready enough to entertain the idea of her equal guilt and her co-operation. He had figured to himself some passionate *hystérique*, merciless as a cat in her hate and her love, a zealous abettor, perhaps even the ruling spirit in the crime.

Then he had seen her, had spoken with her, had helped her in her weakness; and such suspicions, since their first meeting, had seemed the vilest of infamy. He had seen her eyes and her mouth; he had breathed the woman's atmosphere. Trent was one of those who fancy they can scent true wickedness in the air. In her presence he had felt an inward certainty of her ultimate goodness of heart; and it was nothing against this that she had abandoned herself a moment, that day on the cliff, to the sentiment of relief at the ending of her bondage, of her years of starved sympathy and unquickened motherhood. That she had turned to Marlowe in her destitution he believed; that she had any knowledge of his deadly purpose he did not believe.

And yet, morning and evening the sickening doubts returned, and he recalled again that it was almost in her presence that Marlowe had made his preparations in the bedroom of the murdered man, that it was by the window of her own chamber that he had escaped from the house. Had he forgotten his cunning and taken the risk of telling her then? Or had he, as Trent thought more likely, still played his part with her then, and stolen off while she slept? He did not think she had known of the masquerade when she gave evidence at the inquest; it read like honest evidence. Or—the question would never be silenced, though he scorned it—had she lain expecting the footsteps in the room and the whisper that should tell her that it was done? Among the foul possibilities of human nature, was it possible that black ruthlessness and black deceit as well were hidden behind that good and straight and gentle seeming?

These thoughts would scarcely leave him when he was alone.

Trent served Sir James, well earning his pay for six months, and then returned to Paris where he went to work again with a better heart. His powers had returned to him, and he began to live more happily than he had expected among a tribe of strangely assorted friends, French, English, and American, artists, poets, journalists, policemen, hotel-keepers, soldiers, lawyers, business men, and others. His old faculty of sympathetic interest in his fellows won for him, just as in his student days, privileges seldom extended to the Briton. He enjoyed again the rare experience of being taken into the bosom of a Frenchman's family. He was admitted to the momentous confidence of *les jeunes*, and found them as sure that they had surprised the secrets of art and life as the departed *jeunes* of ten years before had been.

The bosom of the Frenchman's family was the same as those he had known in the past, even to the patterns of the wallpaper and movables. But the *jeunes*, he perceived with regret, were totally different from their forerunners. They were much more shallow and puerile, much less really clever. The secrets they wrested from the Universe were not such important and interesting secrets as had been wrested by the old *jeunes*. This he believed and deplored until one day he found himself seated at a restaurant next to a too well-fed man whom, in spite of the ravages of comfortable living, he recognized as one of the *jeunes* of his own period. This one had been wont to describe himself and three or four others as the Hermits of the New Parnassus. He and his school had talked outside cafés and elsewhere more than solitaries do as a rule; but, then, rules were what they had vowed themselves to destroy. They proclaimed that verse, in particular, was free. The Hermit of the New Parnassus was now in the Ministry of the Interior, and already decorated: he expressed to Trent the opinion that what France needed most was a hand of iron. He was able to quote the exact price paid for certain betrayals of the country, of which Trent had not previously heard.

Thus he was brought to make the old discovery that it was he who had changed, like his friend of the Administration, and that *les jeunes* were still the same. Yet he found it hard to say what precisely he had lost that so greatly mattered; unless indeed it were so simple a thing as his high spirits.

One morning in June, as he descended the slope of the Rue des Martyrs, he saw approaching a figure that he remembered. He glanced quickly round, for the thought of meeting Mr Bunner again was unacceptable. For some time he had recognized that his wound was healing under the spell of creative work; he thought less often of the woman he loved, and with less pain. He would not have the memory of those three days reopened.

But the straight and narrow thoroughfare offered no refuge, and the American saw him almost at once.

His unforced geniality made Trent ashamed, for he had liked the man. They sat long over a meal, and Mr Bunner talked. Trent listened to him, now that he was in for it, with genuine pleasure, now and then contributing a question or remark. Besides liking his companion, he enjoyed his conversation, with its unending verbal surprises, for its own sake.

Mr Bunner was, it appeared, resident in Paris as the chief Continental agent of the Manderson firm, and fully satisfied with his position and prospects. He discoursed on these for some twenty minutes. This subject at length exhausted, he went on to tell Trent, who confessed that he had been away from England for a year, that Marlowe had shortly after the death of Manderson entered his father's business, which was now again in a flourishing state, and had already come to be practically in control of it. They had kept up their intimacy, and were even now planning a holiday for the summer. Mr Bunner spoke with generous admiration of his friend's talent for affairs. 'Jack Marlowe has a natural big head,' he declared, 'and if he had more experience, I wouldn't want to have him up against me. He would put a crimp in me every time.'

As the American's talk flowed on, Trent listened with a slowly growing perplexity. It became more and more plain that something was very wrong in his theory of the situation; there was no mention of its central figure. Presently Mr Bunner mentioned that Marlowe was engaged to be married to an Irish girl, whose charms he celebrated with native enthusiasm.

Trent clasped his hands savagely together beneath the table. What could have happened? His ideas were sliding and shifting. At last he forced himself to put a direct question.

Mr Bunner was not very fully informed. He knew that Mrs Manderson had left England immediately after the settlement of her husband's affairs, and had lived for some time in Italy. She had returned not long ago to London, where she had decided not to live in the house in Mayfair, and had bought a smaller one in the Hampstead neighbourhood; also, he understood, one somewhere in the country. She was said to go but little into society. 'And all the good hard dollars just waiting for some one to spraddle them around,' said Mr Bunner, with a note of pathos in his voice. 'Why, she has money to burn—money to feed to the birds—and nothing doing. The old man left her more than half his wad. And think of the figure she might make in the world. She is beautiful, and she is the best woman I ever met, too. But she couldn't ever seem to get the habit of spending money the way it ought to be spent.'

His words now became a soliloquy: Trent's thoughts were occupying all his attention. He pleaded business soon, and the two men parted with cordiality.

Half an hour later Trent was in his studio, swiftly and mechanically 'cleaning up'. He wanted to know what had happened; somehow he must find out. He could never approach herself, he knew; he would never bring back to her the shame of that last encounter with him; it was scarcely likely that he would even set eyes on her. But he must get to know!... Cupples was in London, Marlowe was there.... And, anyhow, he was sick of Paris.

Such thoughts came and went; and below them all strained the fibres of an unseen cord that dragged mercilessly at his heart, and that he cursed bitterly in the moments when he could not deny to himself that it was there. The folly, the useless, pitiable folly of it!

In twenty-four hours his feeble roots in Paris had been torn out. He was looking over a leaden sea at the shining fortress-wall of the Dover cliffs.

But though he had instinctively picked out the lines of a set purpose from among the welter of promptings in his mind, he found it delayed at the very outset.

He had decided that he must first see Mr Cupples, who would be in a position to tell him much more than the American knew. But Mr Cupples was away on his travels, not expected to return for a month; and Trent had no reasonable excuse for hastening his return. Marlowe he would not confront until he had tried at least to reconnoitre the position. He constrained himself not to commit the crowning folly of seeking out Mrs Manderson's house in Hampstead; he could not enter it, and the thought of the possibility of being seen by her lurking in its neighbourhood brought the blood to his face.

He stayed at an hotel, took a studio, and while he awaited Mr Cupples's return attempted vainly to lose himself in work.

At the end of a week he had an idea that he acted upon with eager precipitancy. She had let fall some word at their last meeting, of a taste for music. Trent went that evening, and thenceforward regularly, to the opera. He might see her; and if, in spite of his caution, she caught sight of him, they could be blind to each other's presence—anybody might happen to go to the opera.

So he went alone each evening, passing as quickly as he might through the people in the vestibule; and each evening he came away knowing that she had not been in the house. It was a habit that yielded him a sort of satisfaction along with the guilty excitement of his search; for he too loved music, and nothing gave him so much peace while its magic endured.

One night as he entered, hurrying through the brilliant crowd, he felt a touch on his arm. Flooded with an incredible certainty at the touch, he turned.

It was she: so much more radiant in the absence of grief and anxiety, in the fact that she was smiling, and in the allurement of evening dress, that he could not speak. She, too, breathed a little quickly, and there was a light of daring in her eyes and cheeks as she greeted him.

Her words were few. 'I wouldn't miss a note of *Tristan*,' she said, 'nor must you. Come and see me in the interval.' She gave him the number of the box.

CHAPTER XIII

Eruption

THE following two months were a period in Trent's life that he has never since remembered without shuddering. He met Mrs Manderson half a dozen times, and each time her cool friendliness, a nicely calculated mean between mere acquaintance and the first stage of intimacy, baffled and maddened him. At the opera he had found her, to his further amazement, with a certain Mrs Wallace, a frisky matron whom he had known from childhood. Mrs Manderson, it appeared, on her return from Italy, had somehow wandered into circles to which he belonged by nurture and dis-position. It came, she said, of her having pitched her tent in their hunting-grounds; several of his friends were near neighbours. He had a dim but horrid recollection of having been on that occasion unlike himself, ill at ease, burning in the face, talking with idiot loquacity of his adventures in the Baltic provinces, and finding from time to time that he was addressing himself exclusively to Mrs Wallace. The other lady, when he joined them, had com-pletely lost the slight appearance of agitation with which she had stopped him in the vestibule. She had spoken pleasantly to him of her travels, of her settlement in London, and of people whom they both knew.

During the last half of the opera, which he had stayed in the box to hear, he had been conscious of nothing, as he sat behind them, but the angle of her cheek and the mass of her hair, the lines of her shoulder and arm, her hand upon the cushion. The black hair had seemed at last a forest, immeasurable, pathless and enchanted, luring him to a fatal adventure. . . . At the end he had been pale and sub-dued, parting with them rather formally.

The next time he saw her—it was at a country house where both were guests—and the subsequent times, he had had himself in hand. He had matched her manner and had acquitted himself, he thought, decently, considering—

Considering that he lived in an agony of bewilderment and remorse and longing. He could make nothing, absolutely nothing, of her attitude. That she had read his manuscript and understood the suspicion indicated in his last question to her at White Gables was beyond the possibility of doubt. Then how could she treat him thus amiably and frankly, as she treated all the world of men who had done her no injury?

For it had become clear to his intuitive sense, for all the absence of any shade of differentiation in her outward manner, that an injury had been done, and that she had felt it. Several times, on the rare and brief occasions when they had talked apart, he had warning from the same sense that she was approaching this subject; and each time he had turned the conversation with the ingenuity born of fear. Two resolutions he made. The first was that when he had completed a commissioned work which tied him to London he would go away and stay away. The strain was too great. He no longer burned to know the truth; he wanted nothing to confirm his fixed internal conviction by faith, that he had blundered, that he had misread the situation, misinterpreted her tears, written himself down a slanderous fool. He speculated no more on Marlowe's motive in the killing of Manderson. Mr Cupples returned to London, and Trent asked him nothing. He knew now that he had been right in those words—Trent remembered them for the emphasis with which they were spoken—'So long as she considered herself bound to him . . . no power on earth could have persuaded her.' He met Mrs Manderson at dinner at her uncle's large and tomb-like house in Bloomsbury, and there he conversed most of the evening with a professor of archaeology from Berlin.

His other resolution was that he would not be with her alone.

But when, a few days after, she wrote asking him to come and see her on the following afternoon, he made no attempt to excuse himself. This was a formal challenge.

While she celebrated the rites of tea, and for some little time thereafter, she joined with such natural ease in his slightly fevered conversation on matters of the day that he began to hope she had changed what he could not doubt had been her resolve, to corner him and speak to him gravely. She was to all appearance careless now, smiling so that he recalled, not for the first time since that night at the opera, what was written long ago of a Princess of Brunswick: 'Her mouth has ten thousand charms that touch the soul.' She made a tour of the beautiful room where she had received him, singling out this treasure or that from the spoils of a hundred bric-à-brac shops, laughing over her quests, discoveries, and bargainings. And when he asked if she would delight him again with a favourite piece of his which he had heard her play at another house, she consented at once.

She played with a perfection of execution and feeling that moved him now as it had moved him before. 'You are a musician born,' he said quietly when she had finished, and the last tremor of the music had passed away. 'I knew that before I first heard you.'

'I have played a great deal ever since I can remember. It has been a great comfort to me,' she said simply, and half-turned to him smiling. 'When did you first detect music in me? Oh, of course: I was at the opera. But that wouldn't prove much, would it?'

'No,' he said abstractedly, his sense still busy with the music that had just ended. 'I think I knew it the first time I saw you.' Then understanding of his own words came to him, and turned him rigid. For the first time the past had been invoked.

There was a short silence. Mrs Manderson looked at Trent, then hastily looked away. Colour began to rise in her cheeks, and she pursed her lips as if for whistling. Then with a defiant gesture of the shoulders which he remembered she rose suddenly from the piano and placed herself in a chair opposite to him.

'That speech of yours will do as well as anything,' she began slowly, looking at the point of her shoe, 'to bring us to what I wanted to say. I asked you here today on purpose, Mr Trent, because I couldn't bear it any longer. Ever since the day you left me at White Gables I have been saying to myself that it didn't matter what you thought of me in that affair; that you were certainly not the kind of man to speak to others of what you believed about me, after what you had told me of your reasons for suppressing your manuscript. I asked myself how it could matter. But all the time, of course, I knew it did matter. It mattered horribly. Because what you thought was not true.' She raised her eyes and met his gaze calmly. Trent, with a completely expressionless face, returned her look.

'Since I began to know you,' he said, 'I have ceased to think it.'

'Thank you,' said Mrs Manderson; and blushed suddenly and deeply. Then, playing with a glove, she added, 'But I want you to know what *was* true.

'I did not know if I should ever see you again,' she went on in a lower voice, 'but I felt that if I did I must speak to you about this. I thought it would not be hard to do so, because you seemed to me an understanding person; and besides, a woman who has been married isn't expected to have the same sort of difficulty as a young girl in speaking about such things when it is necessary. And then we did meet again, and I discovered that it was very difficult indeed. You made it difficult.'

'How?' he asked quietly.

'I don't know,' said the lady. 'But yes—I do know. It was just because you treated me exactly as if you had never thought or imagined anything of that sort about me. I had always supposed that if I saw you again you would turn on me that hard, horrible sort of look you had when you asked me that last question—do you remember?— at White Gables. Instead of that you were just like any other acquaintance. You were just'—she hesitated and spread out her hands—'nice. You know. After that first time at the opera when I spoke to you I went home positively wondering if you had really recognized me. I mean, I thought you might have recognized my face without remembering who it was.'

A short laugh broke from Trent in spite of himself, but he said nothing.

She smiled deprecatingly. 'Well, I couldn't remember if you had spoken my name; and I thought it might be so. But the next time, at the Iretons', you did speak it, so I knew; and a dozen times during those few days I almost brought myself to tell you, but never quite. I began to feel that you wouldn't let me, that you would slip away from the subject if I approached it. Wasn't I right? Tell me, please.' He nodded. 'But why?' He remained silent.

'Well,' she said, 'I will finish what I had to say, and then you will tell me, I hope, why you had to make it so hard. When I began to understand that you wouldn't let me talk of the matter to you, it made me more determined than ever. I suppose you didn't realize that I would insist on speaking even if you were quite discouraging. I dare say I couldn't have done it if I had been guilty, as you thought. You walked into my parlour today, never thinking I should dare. Well, now you see.'

Mrs Manderson had lost all her air of hesitancy. She had, as she was wont to say, talked herself enthusiastic, and in the ardour of her purpose to annihilate the misunderstanding that had troubled her so long she felt herself mistress of the situation.

'I am going to tell you the story of the mistake you made,' she continued, as Trent, his hands clasped between his knees, still looked at her enigmatically. 'You will have to believe it, Mr Trent; it is utterly true to life, with its confusions and hidden things and cross-purposes and perfectly natural mistakes that nobody thinks twice about taking for facts. Please understand that I don't blame you in the least, and never did, for jumping to the conclusion you did. You knew that I was estranged from my husband, and you knew what that so often means. You knew before I told you, I expect, that he had taken up an injured attitude towards me; and I was silly enough to try and explain it away. I gave you the explanation of it that I had given myself at first, before I realized the wretched truth; I told you he was disappointed in me because I couldn't take a brilliant lead in society. Well, that was true; he was so. But I could see you weren't convinced. You had guessed what it took me much longer to see, because I knew how irrational it was. Yes; my husband was jealous of John Marlowe; you divined that.

'Then I behaved like a fool when you let me see you had divined it; it was such a blow, you understand, when I had supposed all the humiliation and strain was at an end, and that his delusion had died with him. You practically asked me if my husband's secretary was not my lover, Mr Trent—I *have* to say it, because I want you to understand why I broke down and made a scene. You took that for a confession; you thought I was guilty of that, and I think you even thought I might be a party to the crime, that I had consented. . . . That did hurt me; but perhaps you couldn't have thought anything else—I don't know.'

Trent, who had not hitherto taken his eyes from her face, hung his head at the words. He did not raise it again as she continued. 'But really it was simple shock and distress that made me give way, and the memory of all the misery that mad suspicion had meant to me. And when I pulled myself together again you had gone.'

She rose and went to an escritoire beside the window, unlocked a drawer, and drew out a long, sealed envelope.

'This is the manuscript you left with me,' she said. 'I have read it through again and again. I have always wondered, as everybody does, at your cleverness in things of this kind.' A faintly mischievous smile flashed upon her face, and was gone. 'I thought it was splendid, Mr Trent—I almost forgot that the story was my own, I was so interested. And I want to say now, while I have this in my hand, how much I thank you for your generous, chivalrous act in sacrificing this triumph of yours rather than put a woman's reputation in peril. If all had been as you supposed, the facts must have come out when the police took up the case you put in their hands. Believe me, I understood just what you had done, and I never ceased to be grateful even when I felt most crushed by your suspicion.'

As she spoke her thanks her voice shook a little, and her eyes were bright. Trent perceived nothing of this. His head was still bent. He did not seem to hear. She put the envelope into his hand as it lay open, palm upwards, on his knee. There was a touch of gentleness about the act which made him look up.

'Can you—' he began slowly.

She raised her hand as she stood before him. 'No, Mr Trent; let me finish before you say anything. It is such an unspeakable relief to me to have broken the ice at last, and I want to end the story while I am still feeling the triumph of beginning it.' She sank down into the sofa from which she had first risen. 'I am telling you a thing that nobody else knows. Everybody knew, I suppose, that something had come between us, though I did everything in my power to hide it. But I don't think any one in the world ever guessed what my husband's notion was. People who know me don't think that sort of thing about me, I believe. And his fancy was so ridiculously opposed to the facts. I will tell you what the situation was. Mr Marlowe and I had been friendly enough since he came to us. For all his cleverness—my husband said he had a keener brain than any man he knew—I looked upon him as practically a boy. You know I am a little older than he is, and he had a sort of amiable lack of ambition that made me feel it the more. One day my husband asked me what I thought was the best thing about Marlowe, and not thinking much about it I said, "His manners." He surprised me very much by looking black at that, and after a silence he said, "Yes, Marlowe is a gentleman; that's so", not looking at me.

'Nothing was ever said about that again until about a year ago, when I found that Mr Marlowe had done what I always expected he would do—fallen desperately in love with an American girl. But to my disgust he had picked out the most worthless girl, I do believe, of all those whom we used to meet. She was the daughter of wealthy parents, and she did as she liked with them; very beautiful, well educated, very good at games—what they call a woman-athlete—and caring for nothing on earth but her own amusement. She was one of the most unprincipled flirts I ever knew, and quite the cleverest. Every one knew it, and Mr Marlowe must have heard it; but she made a complete fool of him, brain and all. I don't know how she managed it, but I can imagine. She liked him, of course; but it was quite plain to me that she was playing with him. The whole affair was so idiotic, I got perfectly furious. One day I asked him to row me in a boat on the lake—all this happened at our house by Lake George. We had never been alone together for any length of time before. In the boat I talked to him. I was very kind about it, I think, and he took it admirably, but he didn't believe me a bit. He had the impudence to tell me that I misunderstood Alice's nature. When I hinted at his prospects—I knew he had scarcely anything of his own—he said that if she loved him he could make himself a position in the world. I dare say that was true, with his abilities and his friends—he is rather well connected, you know, as well as popular. But his enlightenment came very soon after that.

'My husband helped me out of the boat when we got back. He joked with Mr Marlowe about something, I remember; for through all that followed he never once changed in his manner to him, and that was one reason why I took so long to realize what he thought about him and myself. But to me he was reserved and silent that evening—not angry. He was always perfectly cold and expressionless to me after he took this idea into his head. After dinner he only spoke to me once. Mr Marlowe was telling him about some horse he had bought for the farm in Kentucky, and my husband looked at me and said, "Marlowe may be a gentleman, but he seldom quits loser in a horse-trade." I was surprised at that, but at that time—and even on the next occasion when he found us together—I didn't understand what was in his mind. That next time was the morning when Mr Marlowe received a sweet little note from the girl asking for his congratulations on her engagement. It was in our New York house. He looked so wretched at breakfast that I thought he was ill, and afterwards I went to the room where he worked, and asked what was the matter. He didn't say anything, but just handed me the note, and turned away to the window. I was very glad that was all over, but terribly sorry for him too, of course. I don't remember what I said, but I remember putting my hand on his arm as he stood there staring out on the garden; and just then my husband appeared at the open door with some papers. He just glanced at us, and then turned and walked quietly back to his study. I thought that he might have heard what I was saying to comfort Mr Marlowe, and that it was rather nice of him to slip away. Mr Marlowe neither saw nor heard him. My husband left the house that morning for the West while I was out. Even then I did not understand. He used often to go off suddenly like that, if some business project called him.

'It was not until he returned a week later that I grasped the situation. He was looking white and strange, and as soon as he saw me he asked me where Mr Marlowe was. Somehow the tone of his question told me everything in a flash.

'I almost gasped; I was wild with indignation. You know, Mr Trent, I don't think I should have minded at all if any one had thought me capable of openly breaking with my husband and leaving him for somebody else. I dare say I might have done that. But that coarse suspicion . . . a man whom he trusted . . . and the notion of conceal-ment. It made me see scarlet. Every shred of pride in me was strung up till I quivered, and I swore to myself on the spot that I would never show by any word or sign that I was conscious of his having such a thought about me. I would behave exactly as I always had behaved, I determined—and that I did, up to the very last. Though I knew that a wall had been made between us now that could never be broken down—even if he asked my pardon and obtained it—I never once showed that I noticed any change.

'And so it went on. I never could go through such a time again. My husband showed silent and cold politeness to me always when we were alone—and that was only when it was unavoidable. He never once alluded to what was in his mind; but I felt it, and he knew that I felt it. Both of us were stubborn in our different attitudes. To Mr Marlowe he was more friendly, if anything, than before—Heaven only knows why. I fancied he was planning some sort of revenge; but that was only a fancy. Certainly Mr Marlowe never knew what was suspected of him. He and I remained good friends, though we never spoke of anything intimate after that disappointment of his; but I made a point of seeing no less of him than I had always done. Then we came to England and to White Gables, and after that followed— my husband's dreadful end.'

She threw out her right hand in a gesture of finality. 'You know about the rest—so much more than any other man,' she added, and glanced up at him with a quaint expression.

Trent wondered at that look, but the wonder was only a passing shadow on his thought. Inwardly his whole being was possessed by thankfulness. All the vivacity had returned to his face. Long before the lady had ended her story he had recognized the certainty of its truth, as from the first days of their renewed acquaintance he had doubted the story that his imagination had built up at White Gables, upon foundations that seemed so good to him.

He said, 'I don't know how to begin the apologies I have to make. There are no words to tell you how ashamed and disgraced I feel when I realize what a crude, cock-sure blundering at a conclusion my suspicion was. Yes, I suspected—you! I had almost forgotten that I was ever such a fool. Almost—not quite. Sometimes when I have been alone I have remembered that folly, and poured contempt on it. I have tried to imagine what the facts were. I have tried to excuse myself.'

She interrupted him quickly. 'What nonsense! Do be sensible, Mr Trent. You had only seen me on two occasions in your life before you came to me with your solution of the mystery.' Again the quaint expression came and was gone. 'If you talk of folly, it really is folly for a man like you to pretend to a woman like me that I had innocence written all over me in large letters—so large that you couldn't believe very strong evidence against me after seeing me twice.'

'What do you mean by "a man like me"?' he demanded with a sort of fierceness. 'Do you take me for a person without any normal instincts? I don't say you impress people as a simple, transparent sort of character—what Mr Calvin Bunner calls a case of open-work; I don't say a stranger might not think you capable of wickedness, if there was good evidence for it: but I say that a man who, after seeing you and being in your atmosphere, could associate you with the particular kind of abomination I imagined, is a fool—the kind of fool who is afraid to trust his senses. . . . As for my making it hard for you to approach the subject, as you say, it is true. It was simply moral cowardice. I understood that you wished to clear the matter up; and

I was revolted at the notion of my injurious blunder being discussed. I tried to show you by my actions that it was as if it had never been. I hoped you would pardon me without any words. I can't forgive myself, and I never shall. And yet if you could know—' He stopped short, and then added quietly, 'Well, will you accept all that as an apology? The very scrubbiest sackcloth made, and the grittiest ashes on the heap. . . . I didn't mean to get worked up,' he ended lamely.

Mrs Manderson laughed, and her laugh carried him away with it. He knew well by this time that sudden rush of cascading notes of mirth, the perfect expression of enjoyment; he had many times tried to amuse her merely for his delight in the sound of it.

'But I love to see you worked up,' she said. 'The bump with which you always come down as soon as you realize that you are up in the air at all is quite delightful. Oh, we're actually both laughing. What a triumphant end to our explanations, after all my dread of the time when I should have it out with you. And now it's all over, and you know; and we'll never speak of it any more.'

'I hope not,' Trent said in sincere relief. 'If you're resolved to be so kind as this about it, I am not high-principled enough to insist on your blasting me with your lightnings. And now, Mrs Manderson, I had better go. Changing the subject after this would be like playing puss-in-the-corner after an earthquake.' He rose to his feet.

'You are right,' she said. 'But no! Wait. There is another thing— part of the same subject; and we ought to pick up all the pieces now while we are about it. Please sit down.' She took the envelope containing Trent's manuscript dispatch from the table where he had laid it. 'I want to speak about this.'

His brows bent, and he looked at her questioningly. 'So do I, if you do,' he said slowly. 'I want very much to know one thing.'

'Tell me.'

'Since my reason for suppressing that information was all a fantasy, why did you never make any use of it? When I began to realize that I had been wrong about you, I explained your silence to myself by saying that you could not bring yourself to do a thing that would put a rope round a man's neck, whatever he might have done. I can quite understand that feeling. Was that what it was? Another possibility I thought of was that you knew of something that was by way of justifying or excusing Marlowe's act. Or I thought you might have a simple horror, quite apart from humanitarian scruples, of appearing publicly in connection with a murder trial. Many important witnesses in such cases have to be practically forced into giving their evidence. They feel there is defilement even in the shadow of the scaffold.'

Mrs Manderson tapped her lips with the envelope without quite concealing a smile. 'You didn't think of another possibility, I suppose, Mr Trent,' she said.

'No.' He looked puzzled.

'I mean the possibility of your having been wrong about Mr Marlowe as well as about me. No, no; you needn't tell me that the chain of evidence is complete. I know it is. But evidence of what? Of Mr Marlowe having impersonated my husband that night, and having escaped by way of my window, and built up an alibi. I have read your dispatch again and again, Mr Trent, and I don't see that those things can be doubted.'

Trent gazed at her with narrowed eyes. He said nothing to fill the brief pause that followed. Mrs Manderson smoothed her skirt with a preoccupied air, as one collecting her ideas.

'I did not make any use of the facts found out by you,' she slowly said at last, 'because it seemed to me very likely that they would be fatal to Mr Marlowe.'

'I agree with you,' Trent remarked in a colourless tone.

'And,' pursued the lady, looking up at him with a mild reasonableness in her eyes, 'as I knew that he was innocent I was not going to expose him to that risk.'

There was another little pause. Trent rubbed his chin, with an affectation of turning over the idea. Inwardly he was telling himself, somewhat feebly, that this was very right and proper; that it was quite feminine, and that he liked her to be feminine. It was permitted to her—more than permitted—to set her loyal belief in the character of a friend above the clearest demonstrations of the intellect. Nevertheless, it chafed him. He would have had her declaration of faith a little less positive in form. It was too irrational to say she 'knew'. In fact (he put it to himself bluntly), it was quite unlike her. If to be unreasonable when reason led to the unpleasant was a specially feminine trait, and if Mrs Manderson had it, she was accustomed to wrap it up better than any woman he had known.

'You suggest,' he said at length, 'that Marlowe constructed an alibi for himself, by means which only a desperate man would have attempted, to clear himself of a crime he did not commit. Did he tell you he was innocent?'

She uttered a little laugh of impatience. 'So you think he has been talking me round. No, that is not so. I am merely sure he did not do it. Ah! I see you think that absurd. But see how unreasonable you are, Mr Trent! Just now you were explaining to me quite sincerely that it was foolishness in you to have a certain suspicion of me after seeing me and being in my atmosphere, as you said.' Trent started in his chair. She glanced at him, and went on: 'Now, I and my atmosphere are much obliged to you, but we must stand up for the rights of other atmospheres. I know a great deal more about Mr Marlowe's atmosphere than you know about mine even now. I saw him constantly for several years. I don't pretend to know all about him; but I do know that he is incapable of a crime of bloodshed. The idea of his planning a murder is as unthinkable to me as the idea of your picking a poor woman's pocket, Mr Trent. I can imagine you killing a man, you know . . . if the man deserved it and had an equal chance of killing you. I could kill a person myself in some circumstances. But Mr Marlowe was incapable of doing it, I don't care what the provocation might be. He had a temper that nothing could shake, and he looked upon human nature with a sort of cold magnanimity that would find excuses for absolutely anything. It wasn't a pose; you could see it was a part of him. He never put it forward, but it was there always. It was quite irritating at times. . . . Now and then in America, I remember, I have heard people talking about lynching, for instance, when he was there. He would sit quite silent and expressionless, appearing not to listen; but you could feel disgust coming from him in waves. He really loathed and hated physical violence. He was a very strange man in some ways, Mr Trent. He gave one a feeling that he might do unexpected things—do you know that feeling one has about some people? What part he really played in the events of that night I have never been able to guess. But nobody who knew anything about him could possibly believe in his deliberately taking a man's life.' Again the movement of her head expressed finality, and she leaned back in the sofa, calmly regarding him.

'Then,' said Trent, who had followed this with earnest attention, 'we are forced back on two other possibilities, which I had not thought worth much consideration until this moment. Accepting what you say, he might still conceivably have killed in self-defence; or he might have done so by accident.'

The lady nodded. 'Of course I thought of those two explanations when I read your manuscript.'

'And I suppose you felt, as I did myself, that in either of those cases the natural thing, and obviously the safest thing, for him to do was to make a public statement of the truth, instead of setting up a series of deceptions which would certainly stamp him as guilty in the eyes of the law, if anything went wrong with them.'

'Yes,' she said wearily, 'I thought over all that until my head ached. And I thought somebody else might have done it, and that he was somehow screening the guilty person. But that seemed wild. I could see no light in the mystery, and after a while I simply let it alone. All I was clear about was that Mr Marlowe was not a murderer, and that if I told what you had found out, the judge and jury would probably think he was. I promised myself that I would speak to you about it if we should meet again; and now I've kept my promise.'

Trent, his chin resting on his hand, was staring at the carpet. The excitement of the hunt for the truth was steadily rising in him. He had not in his own mind accepted Mrs Manderson's account of Marlowe's character as unquestionable. But she had spoken forcibly; he could by no means set it aside, and his theory was much shaken.

'There is only one thing for it,' he said, looking up. 'I must see Marlowe. It worries me too much to have the thing left like this. I will get at the truth. Can you tell me,' he broke off, 'how he behaved after the day I left White Gables?'

'I never saw him after that,' said Mrs Manderson simply. 'For some days after you went away I was ill, and didn't go out of my room. When I got down he had left and was in London, settling things with the lawyers. He did not come down to the funeral. Immediately after that I went abroad. After some weeks a letter from him reached me, saying he had concluded his business and given the solicitors all the assistance in his power. He thanked me very nicely for what he called all my kindness, and said goodbye. There was nothing in it about his plans for the future, and I thought it particularly strange that he said not a word about my husband's death. I didn't answer. Knowing what I knew, I couldn't. In those days I shuddered whenever I thought of that masquerade in the night. I never wanted to see or hear of him again.'

'Then you don't know what has become of him?'

'No; but I dare say Uncle Burton—Mr Cupples, you know—could tell you. Some time ago he told me that he had met Mr Marlowe in London, and had some talk with him. I changed the conversation.' She paused and smiled with a trace of mischief. 'I rather wonder what you supposed had happened to Mr Marlowe after you withdrew from the scene of the drama that you had put together so much to your satisfaction.'

Trent flushed. 'Do you really want to know?' he said.

'I ask you,' she retorted quietly.

'You ask me to humiliate myself again, Mrs Manderson. Very well. I will tell you what I thought I should most likely find when I returned to London after my travels: that you had married Marlowe and gone to live abroad.'

She heard him with unmoved composure. 'We certainly couldn't have lived very comfortably in England on his money and mine,' she observed thoughtfully. 'He had practically nothing then.'

He stared at her—'gaped', she told him some time afterwards. At the moment she laughed with a little embarrassment.

'Dear me, Mr Trent! Have I said anything dreadful? You surely must know. . . . I thought everybody understood by now. . . . I'm sure I've had to explain it often enough . . . if I marry again I lose everything that my husband left me.'

The effect of this speech upon Trent was curious. For an instant his face was flooded with the emotion of surprise. As this passed away he gradually drew himself together, as he sat, into a tense attitude. He looked, she thought as she saw his knuckles grow white on the arms of the chair, like a man prepared for pain under the hand of the surgeon. But all he said, in a voice lower than his usual tone, was, 'I had no idea of it.'

'It is so,' she said calmly, trifling with a ring on her finger. 'Really, Mr Trent, it is not such a very unusual thing. I think I am glad of it. For one thing, it has secured me—at least since it became generally known—from a good many attentions of a kind that a woman in my position has to put up with as a rule.'

'No doubt,' he said gravely. 'And . . . the other kind?'

She looked at him questioningly. 'Ah!' she laughed. 'The

other kind trouble me even less. I have not yet met a man silly
enough to want to marry a widow with a selfish disposition, and
luxurious habits and tastes, and nothing but the little my father left
me.'

She shook her head, and something in the gesture shattered the
last remnants of Trent's self-possession.

'Haven't you, by Heaven!' he exclaimed, rising with a violent
movement and advancing a step towards her. 'Then I am going to
show you that human passion is not always stifled by the smell of
money. I am going to end the business—my business. I am going to
tell you what I dare say scores of better men have wanted to tell you,
but couldn't summon up what I have summoned up—the infernal
cheek to do it. They were afraid of making fools of themselves. I am
not. You have accustomed me to the feeling this afternoon.' He
laughed aloud in his rush of words, and spread out his hands. 'Look at
me! It is the sight of the century! It is one who says he loves you, and
would ask you to give up very great wealth to stand at his side.'

She was hiding her face in her hands. He heard her say brokenly,
'Please . . . don't speak in that way.'

He answered: 'It will make a great difference to me if you will
allow me to say all I have to say before I leave you. Perhaps it is in bad
taste, but I will risk that; I want to relieve my soul; it needs open
confession. This is the truth. You have troubled me ever since the
first time I saw you—and you did not know it—as you sat under the
edge of the cliff at Marlstone, and held out your arms to the sea. It was
only your beauty that filled my mind then. As I passed by you it
seemed as if all the life in the place were crying out a song about you
in the wind and the sunshine. And the song stayed in my ears; but
even your beauty would be no more than an empty memory to me by
now if that had been all. It was when I led you from the hotel there
to your house, with your hand on my arm, that—what was it that
happened? I only knew that your stronger magic had struck home,
and that I never should forget that day, whatever the love of my life
should be. Till that day I had admired as I should admire the loveliness of a still lake; but that day I felt the spell of the divinity of the
lake. And next morning the waters were troubled, and she rose—the
morning when I came to you with my questions, tired out with doubts
that were as bitter as pain, and when I saw you without your pale,
sweet mask of composure—when I saw you moved and glowing, with

your eyes and your hands alive, and when you made me understand that for such a creature as you there had been emptiness and the mere waste of yourself for so long. Madness rose in me then, and my spirit was clamouring to say what I say at last now: that life would never seem a full thing again because you could not love me, that I was taken for ever in the nets of your black hair and by the incantation of your voice—'

'Oh, stop!' she cried, suddenly throwing back her head, her face flaming and her hands clutching the cushions beside her. She spoke fast and disjointedly, her breath coming quick. 'You shall not talk me into forgetting common sense. What does all this mean? Oh, I do not recognize you at all—you seem another man. We are not children; have you forgotten that? You speak like a boy in love for the first time. It is foolish, unreal—I know that if you do not. I will not hear it. What has happened to you?' She was half sobbing. 'How can these sentimentalities come from a man like you? Where is your self-restraint?'

'Gone!' exclaimed Trent, with an abrupt laugh. 'It has got right away. I am going after it in a minute.' He looked gravely down into her eyes. 'I don't care so much now. I never could declare myself to you under the cloud of your great fortune. It was too heavy. There's nothing creditable in that feeling, as I look at it; as a matter of simple fact it was a form of cowardice—fear of what you would think, and very likely say—fear of the world's comment too, I suppose. But the cloud being rolled away, I have spoken, and I don't care so much. I can face things with a quiet mind now that I have told you the truth in its own terms. You may call it sentimentality or any other nickname you like. It is quite true that it was not intended for a scientific statement. Since it annoys you, let it be extinguished. But please believe that it was serious to me if it was comedy to you. I have said that I love you, and honour you, and would hold you dearest of all the world. Now give me leave to go.'

But she held out her hands to him.

CHAPTER XIV

Writing a Letter

'IF you insist,' Trent said, 'I suppose you will have your way. But I had much rather write it when I am not with you. However, if I must, bring me a tablet whiter than a star, or hand of hymning angel; I mean a sheet of note-paper not stamped with your address. Don't under-estimate the sacrifice I am making. I never felt less like correspond-ence in my life.'

She rewarded him.

'What shall I say?' he enquired, his pen hovering over the paper. 'Shall I compare him to a summer's day? What *shall* I say?'

'Say what you want to say,' she suggested helpfully.

He shook his head. 'What I want to say—what I have been wanting for the past twenty-four hours to say to every man, woman, and child I met—is "Mabel and I are betrothed, and all is gas and gaiters." But that wouldn't be a very good opening for a letter of strictly formal, not to say sinister, character. I have got as far as "Dear Mr Marlowe." What comes next?'

'I am sending you a manuscript,' she prompted, 'which I thought you might like to see.'

'Do you realize,' he said, 'that in that sentence there are only two words of more than one syllable? This letter is meant to impress, not to put him at his ease. We must have long words.'

'I don't see why,' she answered. 'I know it is usual, but why is it? I have had a great many letters from lawyers and business people, and they always begin, "with reference to our communication", or some such mouthful, and go on like that all the way through. Yet when I see them they don't talk like that. It seems ridiculous to me.'

'It is not at all ridiculous to them.' Trent laid aside the pen with an appearance of relief and rose to his feet. 'Let me explain. A people like our own, not very fond of using its mind, gets on in the ordinary way with a very small and simple vocabulary. Long words are abnor-mal, and like everything else that is abnormal, they are either very funny or tremendously solemn. Take the phrase "intelligent antici-pation", for instance. If such a phrase had been used in any other country in Europe, it would not have attracted the slightest attention. With us it has become a proverb; we all grin when we hear it in a speech or read it in a leading article; it is considered to be one of the best things ever said. Why? Just because it consists of two long words. The idea expressed is as commonplace as cold mutton. Then there's "terminological inexactitude". How we all roared, and are still roaring, at that! And the whole of the joke is that the words are long. It's just the same when we want to be very serious; we mark it by turning to long words. When a solicitor can begin a sentence with, "pursuant to the instructions communicated to our representative", or some such gibberish, he feels that he is earn-ing his six-and-eightpence. Don't laugh! It is perfectly true. Now Continentals haven't got that feeling. They are always bothering about ideas, and the result is that every shopkeeper or peasant has a vocabulary in daily use that is simply Greek to the vast majority of Britons. I remember some time ago I was dining with a friend of mine who is a Paris cabman. We had dinner at a dirty little res-taurant opposite the central post office, a place where all the clients were cabmen or porters. Conversation was general, and it struck me that a London cabman would have felt a little out of his depth. Words like "functionary" and "unforgettable" and "exterminate" and "independence" hurtled across the table every instant. And these were just ordinary, vulgar, jolly, red-faced cabmen. Mind you,' he went on hurriedly, as the lady crossed the room and took up his pen, 'I merely mention this to illustrate my point. I'm not saying that cab-men ought to be intellectuals. I don't think so; I agree with Keats—happy is England, sweet her artless cabmen, enough their simple loveliness for me. But when you come to the people who make up the collective industrial brain-power of the country. . . . Why, do you know—'

'Oh no, no, no!' cried Mrs Manderson. 'I don't know anything at the moment, except that your talking must be stopped somehow, if we are to get any further with that letter to Mr Marlowe. You shall not get out of it. Come!' She put the pen into his hand.

Trent looked at it with distaste. 'I warn you not to discourage my talking,' he said dejectedly. 'Believe me, men who don't talk are even worse to live with than men who do. O have a care of natures that are mute. I confess I'm shirking writing this thing. It is almost an indecency. It's mixing two moods to write the sort of letter I mean to write, and at the same time to be sitting in the same room with you.'

She led him to his abandoned chair before the escritoire and pushed him gently into it. 'Well, but please try. I want to see what you write, and I want it to go to him at once. You see, I would be contented enough to leave things as they are; but you say you must get at the truth, and if you must, I want it to be as soon as possible. Do it now—you know you can if you will—and I'll send it off the moment it's ready. Don't you ever feel that—the longing to get the worrying letter into the post and off your hands, so that you can't recall it if you would, and it's no use fussing any more about it?'

'I will do as you wish,' he said, and turned to the paper, which he dated as from his hotel. Mrs Manderson looked down at his bent head with a gentle light in her eyes, and made as if to place a smoothing hand upon his rather untidy crop of hair. But she did not touch it. Going in silence to the piano, she began to play very softly. It was ten minutes before Trent spoke.

'If he chooses to reply that he will say nothing?'

Mrs Manderson looked over her shoulder. 'Of course he dare not take that line. He will speak to prevent you from denouncing him.'

'But I'm not going to do that anyhow. You wouldn't allow it—you said so; besides, I won't if you would. The thing's too doubtful now.'

'But,' she laughed, 'poor Mr Marlowe doesn't know you won't, does he?'

Trent sighed. 'What extraordinary things codes of honour are!' he remarked abstractedly. 'I know that there are things I should do, and never think twice about, which would make you feel disgraced if you did them—such as giving any one who grossly insulted me a black eye, or swearing violently when I barked my shin in a dark room. And now you are calmly recommending me to bluff Marlowe by means of a tacit threat which I don't mean; a thing which hell's most abandoned fiend did never, in the drunkenness of guilt—well, anyhow, I won't do it.' He resumed his writing, and the lady, with an indulgent smile, returned to playing very softly.

In a few minutes more, Trent said: 'At last I am his faithfully. Do you want to see it?' She ran across the twilight room, and turned on a reading lamp beside the escritoire. Then, leaning on his shoulder, she read what follows:

DEAR MR MARLOWE, — You will perhaps remember that we met, under unhappy circumstances, in June of last year at Marlstone.

On that occasion it was my duty, as representing a newspaper, to make an independent investigation of the circumstances of the death of the late Sigsbee Manderson. I did so, and I arrived at certain conclusions. You may learn from the enclosed manuscript, which was originally written as a dispatch for my newspaper, what those conclusions were. For reasons which it is not necessary to state I decided at the last moment not to make them public, or to communicate them to you, and they are known to only two persons beside myself.

At this point Mrs Manderson raised her eyes quickly from the letter. Her dark brows were drawn together. 'Two persons?' she said with a note of enquiry.

'Your uncle is the other. I sought him out last night and told him the whole story. Have you anything against it? I always felt uneasy at keeping it from him as I did, because I had led him to expect I should tell him all I discovered, and my silence looked like mystery-making. Now it is to be cleared up finally, and there is no question of shielding you, I wanted him to know everything. He is a very shrewd adviser, too, in a way of his own; and I should like to have him with me when I see Marlowe. I have a feeling that two heads will be better than one on my side of the interview.'

She sighed. 'Yes, of course, uncle ought to know the truth. I hope there is nobody else at all.' She pressed his hand. 'I so much want all that horror buried—buried deep. I am very happy now, dear, but I shall be happier still when you have satisfied that curious mind of yours and found out everything, and stamped down the earth upon it all.' She continued her reading.

Quite recently, however [the letter went on], facts have come to my knowledge which have led me to change my decision. I do not mean that I shall publish what I discovered, but that I have determined to approach you and ask you for a private statement. If you have anything to say which would place the matter in another light, I can imagine no reason why you should withhold it.

I expect, then, to hear from you when and where I may call upon you; unless you would prefer the interview to take place at my hotel. In either case I desire that Mr Cupples, whom you will remember, and who has read the enclosed document, should be present also.—Faithfully yours,

Philip Trent.

'What a very stiff letter!' she said. 'Now I am sure you couldn't have made it any stiffer in your own rooms.'

Trent slipped the letter and enclosure into a long envelope. 'Yes,' he said, 'I think it will make him sit up suddenly. Now this thing mustn't run any risk of going wrong. It would be best to send a special messenger with orders to deliver it into his own hands. If he's away it oughtn't to be left.'

She nodded. 'I can arrange that. Wait here for a little.'

When Mrs Manderson returned, he was hunting through the music cabinet. She sank on the carpet beside him in a wave of dark brown skirts. 'Tell me something, Philip,' she said.

'If it is among the few things that I know.'

'When you saw uncle last night, did you tell him about—about us?'

'I did not,' he answered. 'I remembered you had said nothing about telling any one. It is for you—isn't it?—to decide whether we take the world into our confidence at once or later on.'

'Then will you tell him?' She looked down at her clasped hands. 'I wish *you* to tell him. Perhaps if you think you will guess why. . . . There! that is settled.' She lifted her eyes again to his, and for a time there was silence between them.

He leaned back at length in the deep chair. 'What a world!' he said. 'Mabel, will you play something on the piano that expresses mere joy, the genuine article, nothing feverish or like thorns under a pot, but joy that has decided in favour of the universe? It's a mood that can't last altogether, so we had better get all we can out of it.'

She went to the instrument and struck a few chords while she thought. Then she began to work with all her soul at the theme in the last movement of the Ninth Symphony which is like the sound of the opening of the gates of Paradise.

CHAPTER XV

Double Cunning

AN old oaken desk with a deep body stood by the window in a room that overlooked St James's Park from a height. The room was large, furnished and decorated by some one who had brought taste to the work; but the hand of the bachelor lay heavy upon it. John Marlowe unlocked the desk and drew a long, stout envelope from the back of the well.

'I understand,' he said to Mr Cupples, 'that you have read this.'

'I read it for the first time two days ago,' replied Mr Cupples, who, seated on a sofa, was peering about the room with a benignant face. 'We have discussed it fully.'

Marlowe turned to Trent. 'There is your manuscript,' he said, laying the envelope on the table. 'I have gone over it three times. I do not believe there is another man who could have got at as much of the truth as you have set down there.'

Trent ignored the compliment. He sat by the table gazing stonily at the fire, his long legs twisted beneath his chair. 'You mean, of course,' he said, drawing the envelope towards him, 'that there is more of the truth to be disclosed now. We are ready to hear you as soon as you like. I expect it will be a long story, and the longer the better, so far as I am concerned; I want to understand thoroughly. What we should both like, I think, is some preliminary account of Manderson and your relations with him. It seemed to me from the first that the character of the dead man must be somehow an element in the business.'

'You were right,' Marlowe answered grimly. He crossed the room and seated himself on a corner of the tall cushion-topped fender. 'I will begin as you suggest.'

'I ought to tell you beforehand,' said Trent, looking him in the eyes, 'that although I am here to listen to you, I have not as yet any reason to doubt the conclusions I have stated here.'

He tapped the envelope. 'It is a defence that you will be putting forward—you understand that?'

'Perfectly.' Marlowe was cool and in complete possession of himself, a man different indeed from the worn-out, nervous being Trent remembered at Marlstone a year and a half ago. His tall, lithe figure was held with the perfection of muscular tone. His brow was candid, his blue eyes were clear, though they still had, as he paused collecting his ideas, the look that had troubled Trent at their first meeting. Only the lines of his mouth showed that he knew himself in a position of difficulty, and meant to face it.

'Sigsbee Manderson was not a man of normal mind,' Marlowe began in his quiet voice. 'Most of the very rich men I met with in America had become so by virtue of abnormal greed, or abnormal industry, or abnormal personal force, or abnormal luck. None of them had remarkable intellects. Manderson delighted too in heaping up wealth; he worked incessantly at it; he was a man of dominant will; he had quite his share of luck; but what made him singular was his brain-power. In his own country they would perhaps tell you that it was his ruthlessness in pursuit of his aims that was his most striking characteristic; but there are hundreds of them who would have carried out his plans with just as little consideration for others if they could have formed the plans.

'I'm not saying Americans aren't clever; they are ten times cleverer than we are, as a nation; but I never met another who showed such a degree of sagacity and foresight, such gifts of memory and mental tenacity, such sheer force of intelligence, as there was behind everything Manderson did in his money-making career. They called him the "Napoleon of Wall Street" often enough in the papers; but few people knew so well as I did how much truth there was in the phrase. He seemed never to forget a fact that might be of use to him, in the first place; and he did systematically with the business facts that concerned him what Napoleon did, as I have read, with military facts. He studied them in special digests which were prepared for him at short intervals, and which he always had at hand, so that he could take up his report on coal or wheat or railways, or whatever it might be, in any unoccupied moment. Then he could make a bolder and cleverer plan than any man of them all. People got to know that Manderson would never do the obvious thing, but they got no further; the thing he did do was almost always a surprise, and much of his success flowed from that. The Street got rattled, as they used to put it, when it was known that the old man was out with his gun, and often his opponents seemed to surrender as easily as Colonel Crockett's coon in the story. The scheme I am going to describe to you would have occupied most men long enough. Manderson could have plotted the whole thing, down to the last detail, while he shaved himself.

'I used to think that his strain of Indian blood, remote as it was, might have something to do with the cunning and ruthlessness of the man. Strangely enough, its existence was unknown to any one but himself and me. It was when he asked me to apply my taste for genealogical work to his own obscure family history that I made the discovery that he had in him a share of the blood of the Iroquois chief Montour and his French wife, a terrible woman who ruled the savage politics of the tribes of the Wilderness two hundred years ago. The Mandersons were active in the fur trade on the Pennsylvanian border in those days, and more than one of them married Indian women. Other Indian blood than Montour's may have descended to Manderson, for all I can say, through previous and subsequent unions; some of the wives' antecedents were quite untraceable, and there were so many generations of pioneering before the whole country was brought under civilization. My researches left me with the idea that there is a very great deal of the aboriginal blood present in the genealogical make-up of the people of America, and that it is very widely spread. The newer families have constantly intermarried with the older, and so many of them had a strain of the native in them— and were often rather proud of it, too, in those days. But Manderson had the idea about the disgracefulness of mixed blood, which grew much stronger, I fancy, with the rise of the negro question after the war. He was thunderstruck at what I told him, and was anxious to conceal it from every soul. Of course I never gave it away while he lived, and I don't think he supposed I would; but I have thought since that his mind took a turn against me from that time onward. It happened about a year before his death.'

'Had Manderson,' asked Mr Cupples, so unexpectedly that the others started, 'any definable religious attitude?'

Marlowe considered a moment. 'None that ever I heard of,' he said. 'Worship and prayer were quite unknown to him, so far as I could see, and I never heard him mention religion. I should doubt if he had any real sense of God at all, or if he was capable of knowing God through the emotions. But I understood that as a child he had had a religious upbringing with a strong moral side to it. His private life was, in the usual limited sense, blameless. He was almost ascetic in his habits, except as to smoking. I lived with him four years without ever knowing him to tell a direct verbal falsehood, constantly as he used to practise deceit in other forms. Can you understand the soul of a man who never hesitated to take steps that would have the effect of hoodwinking people, who would use every trick of the markets to mislead, and who was at the same time scrupulous never to utter a direct lie on the most insignificant matter? Manderson was like that, and he was not the only one. I suppose you might compare the state of mind to that of a soldier who is personally a truthful man, but who will stick at nothing to deceive the enemy. The rules of the game allow it; and the same may be said of business as many business men regard it. Only with them it is always wartime.'

'It is a sad world,' observed Mr Cupples.

'As you say,' Marlowe agreed. 'Now I was saying that one could always take Manderson's word if he gave it in a definite form. The first time I ever heard him utter a downright lie was on the night he died; and hearing it, I believe, saved me from being hanged as his murderer.'

Marlowe stared at the light above his head and Trent moved impatiently in his chair. 'Before we come to that,' he said, 'will you tell us exactly on what footing you were with Manderson during the years you were with him?'

'We were on very good terms from beginning to end,' answered Marlowe. 'Nothing like friendship—he was not a man for making friends—but the best of terms as between a trusted employee and his chief. I went to him as private secretary just after getting my degree at Oxford. I was to have gone into my father's business, where I am now, but my father suggested that I should see the world for a year or two. So I took this secretaryship, which seemed to promise a good deal of varied experience, and I had let the year or two run on to four years before the end came. The offer came to me through the last thing in the world I should have put forward as a qualification for a salaried post, and that was chess.'

At the word Trent struck his hands together with a muttered exclamation. The others looked at him in surprise.

'Chess!' repeated Trent. 'Do you know,' he said, rising and approaching Marlowe, 'what was the first thing I noted about you at our first meeting? It was your eye, Mr Marlowe. I couldn't place it then, but I know now where I had seen your eyes before. They were in the head of no less a man than the great Nikolay Korchagin, with whom I once sat in the same railway carriage for two days. I thought I should never forget the chess eye after that, but I could not put a name to it when I saw it in you. I beg your pardon,' he ended suddenly, resuming his marmoreal attitude in his chair.

'I have played the game from my childhood, and with good players,' said Marlowe simply. 'It is an hereditary gift, if you can call it a gift. At the University I was nearly as good as anybody there, and I gave most of my brains to that and the OUDS and playing about generally. At Oxford, as I dare say you know, inducements to amuse oneself at the expense of one's education are endless, and encouraged by the authorities. Well, one day toward the end of my last term, Dr Munro of Queen's, whom I had never defeated, sent for me. He told me that I played a fairish game of chess. I said it was very good of him to say so. Then he said, "They tell me you hunt, too." I said, "Now and then." He asked, "Is there anything else you can do?" "No," I said, not much liking the tone of the conversation—the old man generally succeeded in putting people's backs up. He grunted fiercely, and then told me that enquiries were being made on behalf of a wealthy American man of business who wanted an English secretary. Manderson was the name, he said. He seemed never to have heard it before, which was quite possible, as he never opened a newspaper and had not slept a night outside the college for thirty years. If I could rub up my spelling—as the old gentleman put it—I might have a good chance for the post, as chess and riding and an Oxford education were the only indispensable points.

'Well, I became Manderson's secretary. For a long time I liked the position greatly. When one is attached to an active American plutocrat in the prime of life one need not have many dull moments. Besides, it made me independent. My father had some serious business reverses about that time, and I was glad to be able to do without an allowance from him. At the end of the first year Manderson doubled my salary. "It's big money," he said, "but I guess I don't lose." You see, by that time I was doing a great deal more than accompany him on horseback in the morning and play chess in the evening, which was mainly what he had required. I was attending to his houses, his farm in Ohio, his shooting in Maine, his horses, his cars, and his yacht. I had become a walking railway-guide and an expert cigar-buyer. I was always learning something.

'Well, now you understand what my position was in regard to Manderson during the last two or three years of my connection with him. It was a happy life for me on the whole. I was busy, my work was varied and interesting; I had time to amuse myself too, and money to spend. At one time I made a fool of myself about a girl, and that was not a happy time; but it taught me to understand the great goodness of Mrs Manderson.' Marlowe inclined his head to Mr Cupples as he said this. 'She may choose to tell you about it. As for her husband, he had never varied in his attitude towards me, in spite of the change that came over him in the last months of his life, as you know. He treated me well and generously in his unsympathetic way, and I never had a feeling that he was less than satisfied with his bargain— that was the sort of footing we lived upon. And it was that continuance of his attitude right up to the end that made the revelation so shocking when I was suddenly shown, on the night on which he met his end, the depth of crazy hatred of myself that was in Manderson's soul.'

The eyes of Trent and Mr Cupples met for an instant.

'You never suspected that he hated you before that time?' asked Trent; and Mr Cupples asked at the same moment, 'To what did you attribute it?'

'I never guessed until that night,' answered Marlowe, 'that he had the smallest ill-feeling toward me. How long it had existed I do not know. I cannot imagine why it was there. I was forced to think, when I considered the thing in those awful days after his death, that it was a case of a madman's delusion, that he believed me to be plotting against him, as they so often do. Some such insane conviction must have been at the root of it. But who can sound the abysses of a lunatic's fancy? Can you imagine the state of mind in which a man dooms himself to death with the object of delivering some one he hates to the hangman?'

Mr Cupples moved sharply in his chair. 'You say Manderson was responsible for his own death?' he asked.

Trent glanced at him with an eye of impatience, and resumed his intent watch upon the face of Marlowe. In the relief of speech it was now less pale and drawn.

'I do say so,' Marlowe answered concisely, and looked his questioner in the face. Mr Cupples nodded.

'Before we proceed to the elucidation of your statement,' observed the old gentleman, in a tone of one discussing a point of abstract science, 'it may be remarked that the state of mind which you attribute to Manderson—'

'Suppose we have the story first,' Trent interrupted, gently laying a hand on Mr Cupples's arm. 'You were telling us,' he went on, turning to Marlowe, 'how things stood between you and Manderson. Now will you tell us the facts of what happened that night?'

Marlowe flushed at the barely perceptible emphasis which Trent laid upon the word 'facts'. He drew himself up.

'Bunner and myself dined with Mr and Mrs Manderson that Sunday evening,' he began, speaking carefully. 'It was just like other dinners at which the four of us had been together. Manderson was taciturn and gloomy, as we had latterly been accustomed to see him. We others kept a conversation going. We rose from the table, I suppose, about nine. Mrs Manderson went to the drawing-room, and Bunner went up to the hotel to see an acquaintance. Manderson asked me to come into the orchard behind the house, saying he wished to have a talk. We paced up and down the pathway there, out of earshot from the house, and Manderson, as he smoked his cigar, spoke to me in his cool, deliberate way. He had never seemed more sane, or more well-disposed to me. He said he wanted me to do him an important service. There was a big thing on. It was a secret affair. Bunner knew nothing of it, and the less I knew the better. He wanted me to do exactly as he directed, and not bother my head about reasons.

'This, I may say, was quite characteristic of Manderson's method of going to work. If at times he required a man to be a mere tool in his hand, he would tell him so. He had used me in the same kind of way a dozen times. I assured him he could rely on me, and said I was ready. "Right now?" he asked. I said of course I was.

'He nodded, and said—I tell you his words as well as I can recollect them—"Well, attend to this. There is a man in England now who is in this thing with me. He was to have left tomorrow for Paris by the noon boat from Southampton to Havre. His name is George Harris—at least that's the name he is going by. Do you remember that name?" "Yes," I said, "when I went up to London a week ago you asked me to book a cabin in that name on the boat that goes tomorrow. I gave you the ticket." "Here it is," he said, producing it from his pocket.

' "Now," Manderson said to me, poking his cigar-butt at me with each sentence in a way he used to have, "George Harris cannot leave England tomorrow. I find I shall want him where he is. And I want Bunner where *he* is. But somebody has got to go by that boat and take certain papers to Paris. Or else my plan is going to fall to pieces. Will you go?" I said, "Certainly. I am here to obey orders."

'He bit his cigar, and said, "That's all right; but these are not just ordinary orders. Not the kind of thing one can ask of a man in the ordinary way of his duty to an employer. The point is this. The deal I am busy with is one in which neither myself nor any one known to be connected with me must appear as yet. That is vital. But these people I am up against know your face as well as they know mine. If my secretary is known in certain quarters to have crossed to Paris at this time and to have interviewed certain people—and that would be known as soon as it happened—then the game is up." He threw away his cigar-end and looked at me questioningly.

'I didn't like it much, but I liked failing Manderson at a pinch still less. I spoke lightly. I said I supposed I should have to conceal my identity, and I would do my best. I told him I used to be pretty good at make-up.

'He nodded in approval. He said, "That's good. I judged you would not let me down." Then he gave me my instructions. "You take the car right now," he said, "and start for Southampton—there's no train that will fit in. You'll be driving all night. Barring accidents, you ought to get there by six in the morning. But whenever you arrive, drive straight to the Bedford Hotel and ask for George Harris. If he's there, tell him you are to go over instead of him, and ask him to telephone me here. It is very important he should know that at the earliest moment possible. But if he isn't there, that means he has got the instructions I wired today, and hasn't gone to Southampton. In that case you don't want to trouble about him any more, but just wait for the boat. You can leave the car at a garage under a fancy name—mine must not be given. See about changing your appearance—I don't care how, so you do it well. Travel by the boat as George Harris.

Let on to be anything you like, but be careful, and don't talk much to anybody. When you arrive, take a room at the Hotel St Petersbourg. You will receive a note or message there, addressed to George Harris, telling you where to take the wallet I shall give you. The wallet is locked, and you want to take good care of it. Have you got that all clear?"

'I repeated the instructions. I asked if I should return from Paris after handing over the wallet. "As soon as you like," he said. "And mind this—whatever happens, don't communicate with me at any stage of the journey. If you don't get the message in Paris at once, just wait until you do—days, if necessary. But not a line of any sort to me. Understand? Now get ready as quick as you can. I'll go with you in the car a little way. Hurry."

'That is, as far as I can remember, the exact substance of what Manderson said to me that night. I went to my room, changed into day clothes, and hastily threw a few necessaries into a kit-bag. My mind was in a whirl, not so much at the nature of the business as at the suddenness of it. I think I remember telling you the last time we met'—he turned to Trent—'that Manderson shared the national fondness for doings things in a story-book style. Other things being equal, he delighted in a bit of mystification and melodrama, and I told myself that this was Manderson all over. I hurried downstairs with my bag and rejoined him in the library. He handed me a stout leather letter-case, about eight inches by six, fastened with a strap with a lock on it. I could just squeeze it into my side-pocket. Then I went to get out the car from the garage behind the house.

'As I was bringing it round to the front a disconcerting thought struck me. I remembered that I had only a few shillings in my pocket.

'For some time past I had been keeping myself very short of cash, and for this reason—which I tell you because it is a vital point, as you will see in a minute. I was living temporarily on borrowed money. I had always been careless about money while I was with Manderson, and being a gregarious animal I had made many friends, some of them belonging to a New York set that had little to do but get rid of the large incomes given them by their parents. Still, I was very well paid, and I was too busy even to attempt to go very far with them in that amusing occupation. I was still well on the right side of the ledger until I began, merely out of curiosity, to play at speculation. It's a very old story—particularly in Wall Street. I thought it was easy; I was lucky at first; I would always be prudent—and so on. Then came the day when I went out of my depth. In one week I was separated from my roll, as Bunner expressed it when I told him; and I owed money too. I had had my lesson. Now in this pass I went to Manderson and told him what I had done and how I stood. He heard me with a very grim smile, and then, with the nearest approach to sympathy I had ever found in him, he advanced me a sum on account of my salary that would clear me. "Don't play the markets any more," was all he said.

'Now on that Sunday night Manderson knew that I was practically without any money in the world. He knew that Bunner knew it too. He may have known that I had even borrowed a little more from Bunner for pocket-money until my next cheque was due, which, owing to my anticipation of my salary, would not have been a large one. Bear this knowledge of Manderson's in mind.

'As soon as I had brought the car round I went into the library and stated the difficulty to Manderson.

'What followed gave me, slight as it was, my first impression of something odd being afoot. As soon as I mentioned the word "expenses" his hand went mechanically to his left hip-pocket, where he always kept a little case containing notes to the value of about a hundred pounds in our money. This was such a rooted habit in him that I was astonished to see him check the movement suddenly. Then, to my greater amazement, he swore under his breath. I had never heard him do this before; but Bunner had told me that of late he had often shown irritation in this way when they were alone. "Has he mislaid his note-case?" was the question that flashed through my mind. But it seemed to me that it could not affect his plan at all, and I will tell you why. The week before, when I had gone up to London to carry out various commissions, including the booking of a berth for Mr George Harris, I had drawn a thousand pounds for Manderson from his bankers, and all, at his request, in notes of small amounts. I did not know what this unusually large sum in cash was for, but I did know that the packets of notes were in his locked desk in the library, or had been earlier in the day, when I had seen him fingering them as he sat at the desk.

'But instead of turning to the desk, Manderson stood looking at me. There was fury in his face, and it was a strange sight to see him gradually master it until his eyes grew cold again. "Wait in the car," he said slowly. "I will get some money." We both went out, and as I was getting into my overcoat in the hall I saw him enter the drawing-room, which, you remember, was on the other side of the entrance hall.

'I stepped out on to the lawn before the house and smoked a cigarette, pacing up and down. I was asking myself again and again where that thousand pounds was; whether it was in the drawing-room; and if so, why. Presently, as I passed one of the drawing-room windows, I noticed Mrs Manderson's shadow on the thin silk curtain. She was standing at her escritoire. The window was open, and as I passed I heard her say, "I have not quite thirty pounds here. Will that be enough?" I did not hear the answer, but next moment Manderson's shadow was mingled with hers, and I heard the chink of money. Then, as he stood by the window, and as I was moving away, these words of his came to my ears—and these at least I can repeat exactly, for astonishment stamped them on my memory—"I'm going out now. Marlowe has persuaded me to go for a moonlight run in the car. He is very urgent about it. He says it will help me to sleep, and I guess he is right."

'I have told you that in the course of four years I had never once heard Manderson utter a direct lie about anything, great or small. I believed that I understood the man's queer, skin-deep morality, and I could have sworn that if he was firmly pressed with a question that could not be evaded he would either refuse to answer or tell the truth. But what had I just heard? No answer to any question. A voluntary statement, precise in terms, that was utterly false. The unimaginable had happened. It was almost as if some one I knew well, in a moment of closest sympathy, had suddenly struck me in the face. The blood rushed to my head, and I stood still on the grass. I stood there until I heard his step at the front door, and then I pulled myself together and stepped quickly to the car. He handed me a banker's paper bag with gold and notes in it. "There's more than you'll want there," he said, and I pocketed it mechanically.

'For a minute or so I stood discussing with Manderson—it was by one of those *tours de force* of which one's mind is capable under great excitement—certain points about the route of the long drive before me. I had made the run several times by day, and I believe I spoke quite calmly and naturally about it. But while I spoke my mind was seething in a flood of suddenly born suspicion and fear. I did not know what I feared. I simply felt fear, somehow—I did not know how—connected with Manderson. My soul once opened to it, fear rushed in like an assaulting army. I felt—I knew—that something was altogether wrong and sinister, and I felt myself to be the object of it. Yet Manderson was surely no enemy of mine. Then my thoughts reached out wildly for an answer to the question why he had told that lie. And all the time the blood hammered in my ears, "Where is that money?" Reason struggled hard to set up the suggestion that the two things were not necessarily connected. The instinct of a man in danger would not listen to it. As we started, and the car took the curve into the road, it was merely the unconscious part of me that steered and controlled it, and that made occasional empty remarks as we slid along in the moonlight. Within me was a confusion and vague alarm that was far worse than any definite terror I ever felt.

'About a mile from the house, you remember, one passed on one's left a gate, on the other side of which was the golf-course. There Manderson said he would get down, and I stopped the car. "You've got it all clear?" he asked. With a sort of wrench I forced myself to remember and repeat the directions given me. "That's OK," he said. "Goodbye, then. Stay with that wallet." Those were the last words I heard him speak, as the car moved gently away from him.'

Marlowe rose from his chair and pressed his hands to his eyes. He was flushed with the excitement of his own narrative, and there was in his look a horror of recollection that held both the listeners silent. He shook himself with a movement like a dog's, and then, his hands behind him, stood erect before the fire as he continued his tale.

'I expect you both know what the back-reflector of a motor car is.'

Trent nodded quickly, his face alive with anticipation; but Mr Cupples, who cherished a mild but obstinate prejudice against motor cars, readily confessed to ignorance.

'It is a small round or more often rectangular mirror,' Marlowe explained, 'rigged out from the right side of the screen in front of the driver, and adjusted in such a way that he can see, without turning round, if anything is coming up behind to pass him. It is quite an ordinary appliance, and there was one on this car. As the car moved on, and Manderson ceased speaking behind me, I saw in that mirror a thing that I wish I could forget.'

Marlowe was silent for a moment, staring at the wall before him.

'Manderson's face,' he said in a low tone. 'He was standing in the road, looking after me, only a few yards behind, and the moonlight was full on his face. The mirror happened to catch it for an instant.

'Physical habit is a wonderful thing. I did not shift hand or foot on the controlling mechanism of the car. Indeed, I dare say it steadied me against the shock to have myself braced to the business of driving. You have read in books, no doubt, of hell looking out of a man's eyes, but perhaps you don't know what a good metaphor that is. If I had not known Manderson was there, I should not have recognized the face. It was that of a madman, distorted, hideous in the imbecility of hate, the teeth bared in a simian grin of ferocity and triumph; the eyes. . . . In the little mirror I had this glimpse of the face alone. I saw nothing of whatever gesture there may have been as that writhing white mask glared after me. And I saw it only for a flash. The car went on, gathering speed, and as it went, my brain, suddenly purged of the vapours of doubt and perplexity, was as busy as the throbbing engine before my feet. I knew.

'You say something in that manuscript of yours, Mr Trent, about the swift automatic way in which one's ideas arrange themselves about some new illuminating thought. It is quite true. The awful intensity of ill-will that had flamed after me from those straining eyeballs had poured over my mind like a searchlight. I was thinking quite clearly now, and almost coldly, for I knew what—at least I knew whom—I had to fear, and instinct warned me that it was not a time to give room to the emotions that were fighting to possess me. The man hated me insanely. That incredible fact I suddenly knew. But the face had told me, it would have told anybody, more than that. It was a face of hatred gratified, it proclaimed some damnable triumph. It had gloated over me driving away to my fate. This too was plain to me. And to what fate?

'I stopped the car. It had gone about two hundred and fifty yards, and a sharp bend of the road hid the spot where I had set Manderson down. I lay back in the seat and thought it out. Something was to happen to me. In Paris? Probably—why else should I be sent there, with money and a ticket? But why Paris? That puzzled me, for I had no melodramatic ideas about Paris. I put the point aside for a moment. I turned to the other things that had roused my attention that evening. The lie about my "persuading him to go for a moonlight run". What was the intention of that? Manderson, I said to myself, will be returning without me while I am on my way to Southampton. What will he tell them about me? How account for his returning alone, and without the car? As I asked myself that sinister question there rushed into my mind the last of my difficulties: "Where are the thousand pounds?" And in the same instant came the answer: "The thousand pounds are in my pocket."

'I got up and stepped from the car. My knees trembled and I felt very sick. I saw the plot now, as I thought. The whole of the story about the papers and the necessity of their being taken to Paris was a blind. With Manderson's money about me, of which he would declare I had robbed him, I was, to all appearance, attempting to escape from England, with every precaution that guilt could suggest. He would communicate with the police at once, and would know how to put them on my track. I should be arrested in Paris, if I got so far, living under a false name, after having left the car under a false name, disguised myself, and travelled in a cabin which I had booked in advance, also under a false name. It would be plainly the crime of a man without money, and for some reason desperately in want of it. As for my account of the affair, it would be too preposterous.

'As this ghastly array of incriminating circumstances rose up before me, I dragged the stout letter-case from my pocket. In the intensity of the moment, I never entertained the faintest doubt that I was right, and that the money was there. It would easily hold the packets of notes. But as I felt it and weighed it in my hands it seemed to me there must be more than this. It was too bulky. What more was to be laid to my charge? After all, a thousand pounds was not much to tempt a man like myself to run the risk of penal servitude. In this new agitation, scarcely knowing what I did, I caught the surrounding strap in my fingers just above the fastening and tore the staple out of the lock. Those locks, you know, are pretty flimsy as a rule.'

Here Marlowe paused and walked to the oaken desk before the window. Opening a drawer full of miscellaneous objects, he took out a box of odd keys, and selected a small one distinguished by a piece of pink tape.

He handed it to Trent. 'I keep that by me as a sort of morbid memento. It is the key to the lock I smashed. I might have saved myself the trouble, if I had known that this key was at that moment in the left-hand side-pocket of my overcoat. Manderson must have slipped it in, either while the coat was hanging in the hall or while he sat at my side in the car. I might not have found the tiny thing there for weeks: as a matter of fact I did find it two days after Manderson was dead, but a police search would have found it in five minutes.

And then I—I with the case and its contents in my pocket, my false name and my sham spectacles and the rest of it—I should have had no explanation to offer but the highly convincing one that I didn't know the key was there.'

Trent dangled the key by its tape idly. Then: 'How do you know this is the key of that case?' he asked quickly.

'I tried it. As soon as I found it I went up and fitted it to the lock. I knew where I had left the thing. So do you, I think, Mr Trent. Don't you?' There was a faint shade of mockery in Marlowe's voice.

'*Touché*,' Trent said, with a dry smile. 'I found a large empty letter-case with a burst lock lying with other odds and ends on the dressing-table in Manderson's room. Your statement is that you put it there. I could make nothing of it.' He closed his lips.

'There was no reason for hiding it,' said Marlowe. 'But to get back to my story. I burst the lock of the strap. I opened the case before one of the lamps of the car. The first thing I found in it I ought to have expected, of course, but I hadn't.' He paused and glanced at Trent.

'It was—' began Trent mechanically, and then stopped himself. 'Try not to bring me in any more, if you don't mind,' he said, meeting the other's eye. 'I have complimented you already in that document on your cleverness. You need not prove it by making the judge help you out with your evidence.'

'All right,' agreed Marlowe. 'I couldn't resist just that much. If *you* had been in my place you would have known before I did that Manderson's little pocket-case was there. As soon as I saw it, of course, I remembered his not having had it about him when I asked for money, and his surprising anger. He had made a false step. He had already fastened his note-case up with the rest of what was to figure as my plunder, and placed it in my hands. I opened it. It contained a few notes as usual, I didn't count them.

'Tucked into the flaps of the big case in packets were the other notes, just as I had brought them from London. And with them were two small wash-leather bags, the look of which I knew well. My heart jumped sickeningly again, for this, too, was utterly unexpected. In those bags Manderson kept the diamonds in which he had been investing for some time past. I didn't open them; I could feel the tiny stones shifting under the pressure of my fingers. How many thousands of pounds' worth there were there I have no idea. We had regarded Manderson's diamond-buying as merely a speculative fad. I believe now that it was the earliest movement in the scheme for my ruin. For any one like myself to be represented as having robbed him, there ought to be a strong inducement shown. That had been provided with a vengeance.

'Now, I thought, I have the whole thing plain, and I must act. I saw instantly what I must do. I had left Manderson about a mile from the house. It would take him twenty minutes, fifteen if he walked fast, to get back to the house, where he would, of course, immediately tell his story of robbery, and probably telephone at once to the police in Bishopsbridge. I had left him only five or six minutes ago; for all that I have just told you was as quick thinking as I ever did. It would be easy to overtake him in the car before he neared the house. There would be an awkward interview. I set my teeth as I thought of it, and all my fears vanished as I began to savour the gratification of telling him my opinion of him. There are probably few people who ever positively looked forward to an awkward interview with Manderson; but I was mad with rage. My honour and my liberty had been plotted against with detestable treachery. I did not consider what would follow the interview. That would arrange itself.

'I had started and turned the car, I was already going fast toward White Gables, when I heard the sound of a shot in front of me, to the right.

'Instantly I stopped the car. My first wild thought was that Manderson was shooting at me. Then I realized that the noise had not been close at hand. I could see nobody on the road, though the moonlight flooded it. I had left Manderson at a spot just round the corner that was now about a hundred yards ahead of me. After half a minute or so, I started again, and turned the corner at a slow pace. Then I stopped again with a jar, and for a moment I sat perfectly still.

'Manderson lay dead a few steps from me on the turf within the gate, clearly visible to me in the moonlight.'

Marlowe made another pause, and Trent, with a puckered brow, enquired, 'On the golf-course?'

'Obviously,' remarked Mr Cupples. 'The eighth green is just there.' He had grown more and more interested as Marlowe went on, and was now playing feverishly with his thin beard.

'On the green, quite close to the flag,' said Marlowe. 'He lay on his back, his arms were stretched abroad, his jacket and heavy overcoat were open; the light shone hideously on his white face and his shirt-front; it glistened on his bared teeth and one of the eyes. The other . . . you saw it. The man was certainly dead. As I sat there stunned, unable for the moment to think at all, I could even see a thin dark line of blood running down from the shattered socket to the ear. Close by lay his soft black hat, and at his feet a pistol.

'I suppose it was only a few seconds that I sat helplessly staring at the body. Then I rose and moved to it with dragging feet; for now the truth had come to me at last, and I realized the fullness of my appalling danger. It was not only my liberty or my honour that the maniac had undermined. It was death that he had planned for me; death with the degradation of the scaffold. To strike me down with certainty, he had not hesitated to end his life; a life which was, no doubt, already threatened by a melancholic impulse to self-destruction; and the last agony of the suicide had been turned, perhaps, to a devilish joy by the thought that he dragged down my life with his. For as far as I could see at the moment my situation was utterly hopeless. If it had been desperate on the assumption that Manderson meant to denounce me as a thief, what was it now that his corpse denounced me as a murderer?

'I picked up the revolver and saw, almost without emotion, that it was my own. Manderson had taken it from my room, I suppose, while I was getting out the car. At the same moment I remembered that it was by Manderson's suggestion that I had had it engraved with my initials, to distinguish it from a precisely similar weapon which he had of his own.

'I bent over the body and satisfied myself that there was no life left in it. I must tell you here that I did not notice, then or afterwards, the scratches and marks on the wrists, which were taken as evidence of a struggle with an assailant. But I have no doubt that Manderson deliberately injured himself in this way before firing the shot; it was a part of his plan.

'Though I never perceived that detail, however, it was evident enough as I looked at the body that Manderson had not forgotten, in his last act on earth, to tie me tighter by putting out of court the question of suicide. He had clearly been at pains to hold the pistol at arm's length, and there was not a trace of smoke or of burning on the face. The wound was absolutely clean, and was already ceasing to bleed outwardly. I rose and paced the green, reckoning up the points in the crushing case against me.

'I was the last to be seen with Manderson. I had persuaded him—so he had lied to his wife and, as I afterwards knew, to the butler—to go with me for the drive from which he never returned. My pistol had killed him. It was true that by discovering his plot I had saved myself from heaping up further incriminating facts—flight, concealment, the possession of the treasure. But what need of them, after all? As I stood, what hope was there? What could I do?'

Marlowe came to the table and leaned forward with his hands upon it. 'I want,' he said very earnestly, 'to try to make you understand what was in my mind when I decided to do what I did. I hope you won't be bored, because I must do it. You may both have thought I acted like a fool. But after all the police never suspected me. I walked that green for a quarter of an hour, I suppose, thinking the thing out like a game of chess. I had to think ahead and think coolly; for my safety depended on upsetting the plans of one of the longest-headed men who ever lived. And remember that, for all I knew, there were details of the scheme still hidden from me, waiting to crush me.

'Two plain courses presented themselves at once. Either of them, I thought, would certainly prove fatal. I could, in the first place, do the completely straightforward thing: take back the dead man, tell my story, hand over the notes and diamonds, and trust to the saving power of truth and innocence. I could have laughed as I thought of it. I saw myself bringing home the corpse and giving an account of myself, boggling with sheer shame over the absurdity of my wholly unsupported tale, as I brought a charge of mad hatred and fiendish treachery against a man who had never, as far as I knew, had a word to say against me. At every turn the cunning of Manderson had forestalled me. His careful concealment of such a hatred was a characteristic feature of the stratagem; only a man of his iron self-restraint could have done it. You can see for yourselves how every fact in my statement would appear, in the shadow of Manderson's death, a clumsy lie. I tried to imagine myself telling such a story to the counsel for my defence. I could see the face with which he would listen to it; I could read in the lines of it his thought, that to put forward such an impudent farrago would mean merely the disappearance of any chance there might be of a commutation of the capital sentence.

'True, I had not fled. I had brought back the body; I had handed over the property. But how did that help me? It would only suggest that I had yielded to a sudden funk after killing my man, and had no nerve left to clutch at the fruits of the crime; it would suggest, perhaps, that I had not set out to kill but only to threaten, and that when I found that I had done murder the heart went out of me. Turn it which way I would, I could see no hope of escape by this plan of action.

'The second of the obvious things that I might do was to take the hint offered by the situation, and to fly at once. That too must prove fatal. There was the body. I had no time to hide it in such a way that it would not be found at the first systematic search. But whatever I should do with the body, Manderson's not returning to the house would cause uneasiness in two or three hours at most. Martin would suspect an accident to the car, and would telephone to the police. At daybreak the roads would be scoured and enquiries telegraphed in every direction. The police would act on the possibility of there being foul play. They would spread their nets with energy in such a big business as the disappearance of Manderson. Ports and railway termini would be watched. Within twenty-four hours the body would be found, and the whole country would be on the alert for me—all Europe, scarcely less; I did not believe there was a spot in Christendom where the man accused of Manderson's murder could pass unchallenged, with every newspaper crying the fact of his death into the ears of all the world. Every stranger would be suspect; every man, woman, and child would be a detective. The car, wherever I should abandon it, would put people on my track. If I had to choose between two utterly hopeless courses, I decided, I would take that of telling the preposterous truth.

'But now I cast about desperately for some tale that would seem more plausible than the truth. Could I save my neck by a lie? One after another came into my mind; I need not trouble to remember them now. Each had its own futilities and perils; but every one split upon the fact—or what would be taken for fact—that I had induced Manderson to go out with me, and the fact that he had never returned alive. Notion after notion I swiftly rejected as I paced there by the dead man, and doom seemed to settle down upon me more heavily as the moments passed. Then a strange thought came to me.

'Several times I had repeated to myself half-consciously, as a sort of refrain, the words in which I had heard Manderson tell his wife that I had induced him to go out. "Marlowe has persuaded me to go for a moonlight run in the car. He is very urgent about it." All at once it struck me that, without meaning to do so, I was saying this in Manderson's voice.

'As you found out for yourself, Mr Trent, I have a natural gift of mimicry. I had imitated Manderson's voice many times so success-fully as to deceive even Bunner, who had been much more in his company than his own wife. It was, you remember'—Marlowe turned to Mr Cupples—'a strong, metallic voice, of great carrying power, so unusual as to make it a very fascinating voice to imitate, and at the same time very easy. I said the words carefully to myself again, like this—' he uttered them, and Mr Cupples opened his eyes in amaze-ment—'and then I struck my hand upon the low wall beside me. "Manderson never returned alive?" I said aloud. "But Manderson *shall* return alive!" '

'In thirty seconds the bare outline of the plan was complete in my mind. I did not wait to think over details. Every instant was precious now. I lifted the body and laid it on the floor of the car, covered with a rug. I took the hat and the revolver. Not one trace remained on the green, I believe, of that night's work. As I drove back to White Gables my design took shape before me with a rapidity and ease that filled me with a wild excitement. I should escape yet! It was all so easy if I kept my pluck. Putting aside the unusual and unlikely, I should not fail. I wanted to shout, to scream!

'Nearing the house I slackened speed, and carefully reconnoitred the road. Nothing was moving. I turned the car into the open field on the other side of the road, about twenty paces short of the little door at the extreme corner of the grounds. I brought it to rest behind a stack. When, with Manderson's hat on my head and the pistol in my pocket, I had staggered with the body across the moonlit road and through that door, I left much of my apprehension behind me. With swift action and an unbroken nerve I thought I ought to succeed.'

With a long sigh Marlowe threw himself into one of the deep chairs at the fireside and passed his handkerchief over his damp forehead. Each of his hearers, too, drew a deep breath, but not audibly.

'Everything else you know,' he said. He took a cigarette from a box beside him and lighted it. Trent watched the very slight quiver of the hand that held the match, and privately noted that his own was at the moment not so steady.

'The shoes that betrayed me to you,' pursued Marlowe after a short silence, 'were painful all the time I wore them, but I never dreamed that they had given anywhere. I knew that no footstep of mine must appear by any accident in the soft ground about the hut where I laid the body, or between the hut and the house, so I took the shoes off and crammed my feet into them as soon as I was inside the little door. I left my own shoes, with my own jacket and overcoat, near the body, ready to be resumed later. I made a clear footmark on the soft gravel outside the French window, and several on the drugget round the carpet. The stripping off of the outer clothing of the body, and the dressing of it afterwards in the brown suit and shoes, and putting the things into the pockets, was a horrible business; and getting the teeth out of the mouth was worse. The head—but you don't want to hear about it. I didn't feel it much at the time. I was wriggling my own head out of a noose, you see. I wish I had thought of pulling down the cuffs, and had tied the shoes more neatly. And putting the watch in the wrong pocket was a bad mistake. It had all to be done so hurriedly.

'You were wrong, by the way, about the whisky. After one stiffish drink I had no more; but I filled up a flask that was in the cupboard, and pocketed it. I had a night of peculiar anxiety and effort in front of me, and I didn't know how I should stand it. I had to take some once or twice during the drive. Speaking of that, you give rather a generous allowance of time in your document for doing that run by night. You say that to get to Southampton by half-past six in that car, under the conditions, a man must, even if he drove like a demon, have left Marlstone by twelve at latest. I had not got the body dressed in the other suit, with tie and watch-chain and so forth, until nearly ten minutes past; and then I had to get to the car and start it going. But then I don't suppose any other man would have taken the risks I did in that car at night, without a headlight. It turns me cold to think of it now.

'There's nothing much to say about what I did in the house. I spent the time after Martin had left me in carefully thinking over the remaining steps in my plan, while I unloaded and thoroughly cleaned the revolver, using my handkerchief and a penholder from the desk. I also placed the packets of notes, the note-case, and the diamonds in the roll-top desk, which I opened and relocked with Manderson's key. When I went upstairs it was a trying moment, for though I was safe from the eyes of Martin, as he sat in his pantry, there was a faint possibility of somebody being about on the bedroom floor. I had

sometimes found the French maid wandering about there when the
other servants were in bed. Bunner, I knew, was a deep sleeper. Mrs
Manderson, I had gathered from things I had heard her say, was
usually asleep by eleven; I had thought it possible that her gift of
sleep had helped her to retain all her beauty and vitality in spite of a
marriage which we all knew was an unhappy one. Still it was uneasy
work mounting the stairs, and holding myself ready to retreat to the
library again at the least sound from above. But nothing happened.

'The first thing I did on reaching the corridor was to enter my room
and put the revolver and cartridges back in the case. Then I turned
off the light and went quietly into Manderson's room.

'What I had to do there you know. I had to take off the shoes and
put them outside the door, leave Manderson's jacket, waistcoat,
trousers, and black tie, after taking everything out of the pockets,
select a suit and tie and shoes for the body, and place the dental plate
in the bowl, which I moved from the washing-stand to the bedside,
leaving those ruinous finger-marks as I did so. The marks on the
drawer must have been made when I shut it after taking out the tie.
Then I had to lie down in the bed and tumble it. You know all about
it—all except my state of mind, which you couldn't imagine and I
couldn't describe.

'The worst came when I had hardly begun my operations: the
moment when Mrs Manderson spoke from the room where I supposed her asleep. I was prepared for it happening; it was a possibility;
but I nearly lost my nerve all the same. However. . . .

'By the way, I may tell you this: in the extremely unlikely contin-
gency of Mrs Manderson remaining awake, and so putting out of the
question my escape by way of her window, I had planned simply to
remain where I was a few hours, and then, not speaking to her, to
leave the house quickly and quietly by the ordinary way. Martin
would have been in bed by that time. I might have been heard to
leave, but not seen. I should have done just as I had planned with the
body, and then made the best time I could in the car to Southampton.
The difference would have been that I couldn't have furnished an
unquestionable alibi by turning up at the hotel at 6.30. I should have
made the best of it by driving straight to the docks, and making my
ostentatious enquiries there. I could in any case have got there long
before the boat left at noon. I couldn't see that anybody could suspect
me of the supposed murder in any case; but if any one had, and if I

hadn't arrived until ten o'clock, say, I shouldn't have been able to answer, "It is impossible for me to have got to Southampton so soon after shooting him." I should simply have had to say I was delayed by a breakdown after leaving Manderson at half-past ten, and challenged any one to produce any fact connecting me with the crime. They couldn't have done it. The pistol, left openly in my room, might have been used by anybody, even if it could be proved that that particular pistol was used. Nobody could reasonably connect me with the shooting so long as it was believed that it was Manderson who had returned to the house. The suspicion could not, I was confident, enter any one's mind. All the same, I wanted to introduce the element of absolute physical impossibility; I knew I should feel ten times as safe with that. So when I knew from the sound of her breathing that Mrs Manderson was asleep again, I walked quickly across her room in my stocking feet, and was on the grass with my bundle in ten seconds. I don't think I made the least noise. The curtain before the window was of soft, thick stuff and didn't rustle, and when I pushed the glass doors further open there was not a sound.'

'Tell me,' said Trent, as the other stopped to light a new cigarette, 'why you took the risk of going through Mrs Manderson's room to escape from the house. I could see when I looked into the thing on the spot why it had to be on that side of the house; there was a danger of being seen by Martin, or by some servant at a bedroom window, if you got out by a window on one of the other sides. But there were three unoccupied rooms on that side; two spare bedrooms and Mrs Manderson's sitting-room. I should have thought it would have been safer, after you had done what was necessary to your plan in Manderson's room, to leave it quietly and escape through one of those three rooms. . . . The fact that you went through her window, you know,' he added coldly, 'would have suggested, if it became known, various suspicions in regard to the lady herself. I think you understand me.'

Marlowe turned upon him with a glowing face. 'And I think you will understand me, Mr Trent,' he said in a voice that shook a little, 'when I say that if such a possibility had occurred to me then, I would have taken any risk rather than make my escape by that way. . . . Oh well!' he went on more coolly, 'I suppose that to any one who didn't know her, the idea of her being privy to her husband's murder might not seem so indescribably fatuous. Forgive the expression.' He looked attentively at the burning end of his cigarette, studiously unconscious of the red flag that flew in Trent's eyes for an instant at his words and the tone of them.

That emotion, however, was conquered at once. 'Your remark is perfectly just,' Trent said with answering coolness. 'I can quite believe, too, that at the time you didn't think of the possibility I mentioned. But surely, apart from that, it would have been safer to do as I said; go by the window of an unoccupied room.'

'Do you think so?' said Marlowe. 'All I can say is, I hadn't the nerve to do it. I tell you, when I entered Manderson's room I shut the door of it on more than half my terrors. I had the problem confined before me in a closed space, with only one danger in it, and that a *known* danger: the danger of Mrs Manderson. The thing was almost done; I had only to wait until she was certainly asleep after her few moments of waking up, for which, as I told you, I was prepared as a possibility. Barring accidents, the way was clear. But now suppose that I, carrying Manderson's clothes and shoes, had opened that door again and gone in my shirt-sleeves and socks to enter one of the empty rooms. The moonlight was flooding the corridor through the end window. Even if my face was concealed, nobody could mistake my standing figure for Manderson's. Martin might be going about the house in his silent way. Bunner might come out of his bedroom. One of the servants who were supposed to be in bed might come round the corner from the other passage—I had found Célestine prowling about quite as late as it was then. None of these things was very likely; but they were all too likely for me. They were uncertainties. Shut off from the household in Manderson's room I knew exactly what I had to face. As I lay in my clothes in Manderson's bed and listened for the almost inaudible breathing through the open door, I felt far more ease of mind, terrible as my anxiety was, than I had felt since I saw the dead body on the turf. I even congratulated myself that I had had the chance, through Mrs Manderson's speaking to me, of tightening one of the screws in my scheme by repeating the statement about my having been sent to Southampton.'

Marlowe looked at Trent, who nodded as who should say that his point was met.

'As for Southampton,' pursued Marlowe, 'you know what I did when I got there, I have no doubt. I had decided to take Manderson's story about the mysterious Harris and act it out on my own lines. It was a carefully prepared lie, better than anything I could improvise. I even went so far as to get through a trunk call to the hotel at Southampton from the library before starting, and ask if Harris was there. As I expected, he wasn't.'

'Was that why you telephoned?' Trent enquired quickly.

'The reason for telephoning was to get myself into an attitude in which Martin couldn't see my face or anything but the jacket and hat, yet which was a natural and familiar attitude. But while I was about it, it was obviously better to make a genuine call. If I had simply pretended to be telephoning, the people at the exchange could have told you at once that there hadn't been a call from White Gables that night.'

'One of the first things I did was to make that enquiry,' said Trent. 'That telephone call, and the wire you sent from Southampton to the dead man to say Harris hadn't turned up, and you were returning—I particularly appreciated both those.'

A constrained smile lighted Marlowe's face for a moment. 'I don't know that there's anything more to tell. I returned to Marlstone, and faced your friend the detective with such nerve as I had left. The worst was when I heard you had been put on the case—no, that wasn't the worst. The worst was when I saw you walk out of the shrubbery the next day, coming away from the shed where I had laid the body. For one ghastly moment I thought you were going to give me in charge on the spot. Now I've told you everything, you don't look so terrible.'

He closed his eyes, and there was a short silence. Then Trent got suddenly to his feet.

'Cross-examination?' enquired Marlowe, looking at him gravely.

'Not at all,' said Trent, stretching his long limbs. 'Only stiffness of the legs. I don't want to ask any questions. I believe what you have told us. I don't believe it simply because I always liked your face, or because it saves awkwardness, which are the most usual reasons for believing a person, but because my vanity will have it that no man could lie to me steadily for an hour without my perceiving it. Your story is an extraordinary one; but Manderson was an extraordinary man, and so are you. You acted like a lunatic in doing what you did; but I quite agree with you that if you had acted like a sane man you wouldn't have had the hundredth part of a dog's chance with a judge and jury. One thing is beyond dispute on any reading of the affair: you are a man of courage.'

The colour rushed into Marlowe's face, and he hesitated for words. Before he could speak Mr Cupples arose with a dry cough.

'For my part,' he said, 'I never supposed you guilty for a moment.' Marlowe turned to him in grateful amazement, Trent with an incredulous stare. 'But,' pursued Mr Cupples, holding up his hand, 'there is one question which I should like to put.'

Marlowe bowed, saying nothing.

'Suppose,' said Mr Cupples, 'that some one else had been suspected of the crime and put upon trial. What would you have done?'

'I think my duty was clear. I should have gone with my story to the lawyers for the defence, and put myself in their hands.'

Trent laughed aloud. Now that the thing was over, his spirits were rapidly becoming ungovernable. 'I can see their faces!' he said. 'As a matter of fact, though, nobody else was ever in danger. There wasn't a shred of evidence against any one. I looked up Murch at the Yard this morning, and he told me he had come round to Bunner's view, that it was a case of revenge on the part of some American black-hand gang. So there's the end of the Manderson case. Holy, suffering Moses! *What* an ass a man can make of himself when he thinks he's being preternaturally clever!' He seized the bulky envelope from the table and stuffed it into the heart of the fire. 'There's for you, old friend! For want of you the world's course will not fail. But look here! It's getting late—nearly seven, and Cupples and I have an appointment at half-past. We must go. Mr Marlowe, goodbye.' He looked into the other's eyes. 'I am a man who has worked hard to put a rope round your neck. Considering the circumstances, I don't know whether you will blame me. Will you shake hands?'

CHAPTER XVI

The Last Straw

'WHAT was that you said about our having an appointment at half-past seven?' asked Mr Cupples as the two came out of the great gateway of the pile of flats. 'Have we such an appointment?'

'Certainly we have,' replied Trent. 'You are dining with me. Only one thing can properly celebrate this occasion, and that is a dinner for which I pay. No, no! I asked you first. I have got right down to the bottom of a case that must be unique—a case that has troubled even my mind for over a year—and if that isn't a good reason for standing a dinner, I don't know what is. Cupples, we will not go to my club. This is to be a festival, and to be seen in a London club in a state of pleasurable emotion is more than enough to shatter any man's career. Besides that, the dinner there is always the same, or, at least, they always make it taste the same, I know not how. The eternal dinner at my club hath bored millions of members like me, and shall bore; but tonight let the feast be spread in vain, so far as we are concerned. We will not go where the satraps throng the hall. We will go to Sheppard's.'

'Who is Sheppard?' asked Mr Cupples mildly, as they proceeded up Victoria Street. His companion went with an unnatural lightness, and a policeman, observing his face, smiled indulgently at a look of happiness which he could only attribute to alcohol.

'Who is Sheppard?' echoed Trent with bitter emphasis. 'That question, if you will pardon me for saying so, Cupples, is thoroughly characteristic of the spirit of aimless enquiry prevailing in this restless day. I suggest our dining at Sheppard's, and instantly you fold your arms and demand, in a frenzy of intellectual pride, to know who Sheppard is before you will cross the threshold of Sheppard's. I am not going to pander to the vices of the modern mind. Sheppard's is a place where one can dine. I do not know Sheppard. It never occurred to me that Sheppard existed. Probably he is a myth of totemistic origin. All I know is that you can get a bit of saddle of mutton at

Sheppard's that has made many an American visitor curse the day that Christopher Columbus was born. . . . Taxi!'

A cab rolled smoothly to the kerb, and the driver received his instructions with a majestic nod.

'Another reason I have for suggesting Sheppard's,' continued Trent, feverishly lighting a cigarette, 'is that I am going to be married to the most wonderful woman in the world. I trust the connection of ideas is clear.'

'You are going to marry Mabel!' cried Mr Cupples. 'My dear friend, what good news this is! Shake hands, Trent; this is glorious! I congratulate you both from the bottom of my heart. And may I say—I don't want to interrupt your flow of high spirits, which is very natural indeed, and I remember being just the same in similar circumstances long ago—but may I say how earnestly I have hoped for this? Mabel has seen so much unhappiness, yet she is surely a woman formed in the great purpose of humanity to be the best influence in the life of a good man. But I did not know her mind as regarded yourself. *Your* mind I have known for some time,' Mr Cupples went on, with a twinkle in his eye that would have done credit to the worldliest of creatures. 'I saw it at once when you were both dining at my house, and you sat listening to Professor Peppmüller and looking at her. Some of us older fellows have our wits about us still, my dear boy.'

'Mabel says she knew it before that,' replied Trent, with a slightly crestfallen air. 'And I thought I was acting the part of a person who was not mad about her to the life. Well, I never was any good at dissembling. I shouldn't wonder if even old Peppmüller noticed something through his double convex lenses. But however crazy I may have been as an undeclared suitor,' he went on with a return to vivacity, 'I am going to be much worse now. As for your congratulations, thank you a thousand times, because I know you mean them. You are the sort of uncomfortable brute who would pull a face three feet long if you thought we were making a mistake. By the way, I can't help being an ass tonight; I'm obliged to go on blithering. You must try to bear it. Perhaps it would be easier if I sang you a song—one of your old favourites. What was that song you used always to be singing? Like this, wasn't it?' He accompanied the following stave with a dexterous clog-step on the floor of the cab:

'There was an old nigger, and he had a wooden leg.
He had no tobacco, no tobacco could he beg.
Another old nigger was as cunning as a fox,
And he always had tobacco in his old tobacco-box.

'Now for the chorus!

'Yes, he always had tobacco in his old tobacco-box.

'But you're not singing. I thought you would be making the welkin ring.'

'I never sang that song in my life,' protested Mr Cupples. 'I never heard it before.'

'Are you sure?' enquired Trent doubtfully. 'Well, I suppose I must take your word for it. It is a beautiful song, anyhow: not the whole warbling grove in concert heard can beat it. Somehow it seems to express my feelings at the present moment as nothing else could; it rises unbidden to the lips. Out of the fullness of the heart the mouth speaketh, as the Bishop of Bath and Wells said when listening to a speech of Mr Balfour's.'

'When was that?' asked Mr Cupples.

'On the occasion,' replied Trent, 'of the introduction of the Compulsory Notification of Diseases of Poultry Bill, which ill-fated measure you of course remember. Hullo!' he broke off, as the cab rushed down a side street and swung round a corner into a broad and populous thoroughfare, 'we're there already'. The cab drew up.

'Here we are,' said Trent, as he paid the man, and led Mr Cupples into a long, panelled room set with many tables and filled with a hum of talk. 'This is the house of fulfilment of craving, this is the bower with the roses around it. I see there are three bookmakers eating pork at my favourite table. We will have that one in the opposite corner.'

He conferred earnestly with a waiter, while Mr Cupples, in a pleasant meditation, warmed himself before the great fire. 'The wine here,' Trent resumed, as they seated themselves, 'is almost certainly made out of grapes. What shall we drink?'

Mr Cupples came out of his reverie. 'I think,' he said, 'I will have milk and soda water.'

'Speak lower!' urged Trent. 'The head-waiter has a weak heart, and might hear you. Milk and soda water! Cupples, you may think you have a strong constitution, and I don't say you have not, but I warn you that this habit of mixing drinks has been the death of many a robuster man than you. Be wise in time. Fill high the bowl with Samian wine, leave soda to the Turkish hordes. Here comes our food.' He gave another order to the waiter, who ranged the dishes before them and darted away. Trent was, it seemed, a respected customer. 'I have sent,' he said, 'for wine that I know, and I hope you will try it. If you have taken a vow, then in the name of all the teetotal saints drink water, which stands at your elbow, but don't seek a cheap notoriety by demanding milk and soda.'

'I have never taken any pledge,' said Mr Cupples, examining his mutton with a favourable eye. 'I simply don't care about wine. I bought a bottle once and drank it to see what it was like, and it made me ill. But very likely it was bad wine. I will taste some of yours, as it is your dinner, and I do assure you, my dear Trent, I should like to do something unusual to show how strongly I feel on the present occasion. I have not been so delighted for many years. To think,' he reflected aloud as the waiter filled his glass, 'of the Manderson mystery disposed of, the innocent exculpated, and your own and Mabel's happiness crowned—all coming upon me together! I drink to you, my dear friend.' And Mr Cupples took a very small sip of the wine.

'You have a great nature,' said Trent, much moved. 'Your outward semblance doth belie your soul's immensity. I should have expected as soon to see an elephant conducting at the opera as you drinking my health. Dear Cupples! May his beak retain ever that delicate rose-stain!—No, curse it all!' he broke out, surprising a shade of discomfort that flitted over his companion's face as he tasted the wine again. 'I have no business to meddle with your tastes. I apologize. You shall have what you want, even if it causes the head-waiter to perish in his pride.'

When Mr Cupples had been supplied with his monastic drink, and the waiter had retired, Trent looked across the table with significance. 'In this babble of many conversations,' he said, 'we can speak as freely as if we were on a bare hillside. The waiter is whispering soft nothings into the ear of the young woman at the pay-desk. We are alone. What do you think of that interview of this afternoon?' He began to dine with an appetite.

Without pausing in the task of cutting his mutton into very small pieces Mr Cupples replied: 'The most curious feature of it, in my judgement, was the irony of the situation. We both held the clue to that mad hatred of Manderson's which Marlowe found so mysterious. We knew of his jealous obsession; which knowledge we withheld, as was very proper, if only in consideration of Mabel's feelings. Marlowe will never know of what he was suspected by that person. Strange! Nearly all of us, I venture to think, move unconsciously among a network of opinions, often quite erroneous, which other people entertain about us. I remember, for instance, discovering quite by accident some years ago that a number of people of my acquaintance believed me to have been secretly received into the Church of Rome. This absurd fiction was based upon the fact, which in the eyes of many appeared conclusive, that I had expressed myself in talk as favouring the plan of a weekly abstinence from meat. Manderson's belief in regard to his secretary probably rested upon a much slighter ground. It was Mr Bunner, I think you said, who told you of his rooted and apparently hereditary temper of suspicious jealousy. . . . With regard to Marlowe's story, it appeared to me entirely straightforward, and not, in its essential features, especially remarkable, once we have admitted, as we surely must, that in the case of Manderson we have to deal with a more or less disordered mind.'

Trent laughed loudly. 'I confess,' he said, 'that the affair struck me as a little unusual.'

'Only in the development of the details,' argued Mr Cupples. 'What is there abnormal in the essential facts? A madman conceives a crazy suspicion; he hatches a cunning plot against his fancied injurer; it involves his own destruction. Put thus, what is there that any man with the least knowledge of the ways of lunatics would call remarkable? Turn now to Marlowe's proceedings. He finds himself in a perilous position from which, though he is innocent, telling the truth will not save him. Is that an unheard-of situation? He escapes by means of a bold and ingenious piece of deception. That seems to me a thing that might happen every day, and probably does so.' He attacked his now unrecognizable mutton.

'I should like to know,' said Trent, after an alimentary pause in the conversation, 'whether there is anything that ever happened on the face of the earth that you could not represent as quite ordinary and commonplace by such a line of argument as that.'

A gentle smile illuminated Mr Cupples's face. 'You must not suspect me of empty paradox,' he said. 'My meaning will become clearer, perhaps, if I mention some things which *do* appear to me essentially remarkable. Let me see. . . . Well, I would call the life history of the liver-fluke, which we owe to the researches of Poulton, an essentially remarkable thing.'

'I am unable to argue the point,' replied Trent. 'Fair science may have smiled upon the liver-fluke's humble birth, but I never even heard it mentioned.'

'It is not, perhaps, an appetizing subject,' said Mr Cupples thoughtfully, 'and I will not pursue it. All I mean is, my dear Trent, that there are really remarkable things going on all round us if we will only see them; and we do our perceptions no credit in regarding as remarkable only those affairs which are surrounded with an accumulation of sensational detail.'

Trent applauded heartily with his knife-handle on the table, as Mr Cupples ceased and refreshed himself with milk and soda water. 'I have not heard you go on like this for years,' he said. 'I believe you must be almost as much above yourself as I am. It is a bad case of the unrest which men miscall delight. But much as I enjoy it, I am not going to sit still and hear the Manderson affair dismissed as commonplace. You may say what you like, but the idea of impersonating Manderson in those circumstances was an extraordinarily ingenious idea.'

'Ingenious—certainly!' replied Mr Cupples. 'Extraordinarily so—no! In those circumstances (your own words) it was really not strange that it should occur to a clever man. It lay almost on the surface of the situation. Marlowe was famous for his imitation of Manderson's voice; he had a talent for acting; he had a chess-player's mind; he knew the ways of the establishment intimately. I grant you that the idea was brilliantly carried out; but everything favoured it. As for the essential idea, I do not place it, as regards ingenuity, in the same class with, for example, the idea of utilizing the force of recoil in a discharged firearm to actuate the mechanism of ejecting and reloading. I do, however, admit, as I did at the outset, that in respect of details the case had unusual features. It developed a high degree of complexity.'

'Did it really strike you in that way?' enquired Trent with desperate sarcasm.

'The affair became complicated,' went on Mr Cupples unmoved, 'because after Marlowe's suspicions were awakened, a second subtle mind came in to interfere with the plans of the first. That sort of duel often happens in business and politics, but less frequently, I imagine, in the world of crime.'

'I should say never,' Trent replied; 'and the reason is, that even the cleverest criminals seldom run to strategic subtlety. When they do, they don't get caught, since clever policemen have if possible less strategic subtlety than the ordinary clever criminal. But that rather deep quality seems very rarely to go with the criminal make-up. Look at Crippen. He was a very clever criminal as they go. He solved the central problem of every clandestine murder, the disposal of the body, with extreme neatness. But how far did he see through the game? The criminal and the policeman are often swift and bold tacticians, but neither of them is good for more than a quite simple plan. After all, it's a rare faculty in any walk of life.'

'One disturbing reflection was left on my mind,' said Mr Cupples, who seemed to have had enough of abstractions for the moment, 'by what we learned today. If Marlowe had suspected nothing and walked into the trap, he would almost certainly have been hanged. Now how often may not a plan to throw the guilt of murder on an innocent person have been practised successfully? There are, I imagine, numbers of cases in which the accused, being found guilty on circumstantial evidence, have died protesting their innocence. I shall never approve again of a death-sentence imposed in a case decided upon such evidence.'

'I never have done so, for my part,' said Trent. 'To hang in such cases seems to me flying in the face of the perfectly obvious and sound principle expressed in the saying that "you never can tell". I agree with the American jurist who lays it down that we should not hang a yellow dog for stealing jam on circumstantial evidence, not even if he has jam all over his nose. As for attempts being made by malevolent persons to fix crimes upon innocent men, of course it is constantly happening. It's a marked feature, for instance, of all systems of rule by coercion, whether in Ireland or Russia or India or Korea; if the police cannot get hold of a man they think dangerous by fair means, they do it by foul. But there's one case in the State Trials that is peculiarly to the point, because not only was it a case of fastening a murder on innocent people, but the plotter did in effect what Manderson did; he gave up his own life in order to secure the death of his victims. Probably you have heard of the Campden Case.'

Mr Cupples confessed his ignorance and took another potato.

'John Masefield has written a very remarkable play about it,' said Trent, 'and if it ever comes on again in London, you should go and see it, if you like having the fan-tods. I have often seen women weeping in an undemonstrative manner at some slab of oleo-margarine sentiment in the theatre. By George! what everlasting smelling-bottle hysterics they ought to have if they saw that play decently acted! Well, the facts were that John Perry accused his mother and brother of murdering a man, and swore he had helped them to do it. He told a story full of elaborate detail, and had an answer to everything, except the curious fact that the body couldn't be found; but the judge, who was probably drunk at the time—this was in Restoration days—made nothing of that. The mother and brother denied the accusation. All three prisoners were found guilty and hanged, purely on John's evidence. Two years after, the man whom they were hanged for murdering came back to Campden. He had been kidnapped by pirates and taken to sea. His disappearance had given John his idea. The point about John is, that his including himself in the accusation, which amounted to suicide, was the thing in his evidence which convinced everybody of its truth. It was so obvious that no man would do himself to death to get somebody else hanged. Now that is exactly the answer which the prosecution would have made if Marlowe had told the truth. Not one juryman in a million would have believed in the Manderson plot.'

Mr Cupples mused upon this a few moments. 'I have not your acquaintance with that branch of history,' he said at length; 'in fact, I have none at all. But certain recollections of my own childhood return to me in connection with this affair. We know from the things Mabel told you what may be termed the spiritual truth underlying this matter; the insane depth of jealous hatred which Manderson concealed. We can understand that he was capable of such a scheme. But as a rule it is in the task of penetrating to the spiritual truth that the administration of justice breaks down. Sometimes that truth is deliberately concealed, as in Manderson's case. Sometimes, I think, it is concealed because simple people are actually unable to express it, and nobody else divines it. When I was a lad in Edinburgh the whole country went mad about the Sandyford Place murder.'

Trent nodded. 'Mrs M'Lachlan's case. She was innocent right enough.'

'My parents thought so,' said Mr Cupples. 'I thought so myself when I became old enough to read and understand that excessively sordid story. But the mystery of the affair was so dark, and the task of getting at the truth behind the lies told by everybody concerned proved so hopeless, that others were just as fully convinced of the innocence of old James Fleming. All Scotland took sides on the question. It was the subject of debates in Parliament. The press divided into two camps, and raged with a fury I have never seen equalled. Yet it is obvious, is it not?—for I see you have read of the case—that if the spiritual truth about that old man could have been known there would have been very little room for doubt in the matter. If what some surmised about his disposition was true, he was quite capable of murdering Jessie M'Pherson and then casting the blame on the poor feeble-minded creature who came so near to suffering the last penalty of the law.'

'Even a commonplace old dotard like Fleming can be an un-fathomable mystery to all the rest of the human race,' said Trent, 'and most of all in a court of justice. The law certainly does not shine when it comes to a case requiring much delicacy of perception. It goes wrong easily enough over the Flemings of this world. As for the people with temperaments who get mixed up in legal proceedings, they must feel as if they were in a forest of apes, whether they win or lose. Well, I dare say it's good for their sort to have their noses rubbed in reality now and again. But what would twelve red-faced realities in a jury-box have done to Marlowe? His story would, as he says, have been a great deal worse than no defence at all. It's not as if there were a single piece of evidence in support of his tale. Can't you imagine how the prosecution would tear it to rags? Can't you see the judge simply taking it in his stride when it came to the summing up? And the jury—you've served on juries, I expect—in their room, snorting with indignation over the feebleness of the lie, telling each other it was the clearest case they ever heard of, and that they'd have thought better of him if he hadn't lost his nerve at the crisis, and had cleared off with the swag as he intended. Imagine yourself on that jury, not knowing Marlowe, and trembling with indignation at the record un-rolled before you—cupidity, murder, robbery, sudden cowardice, shameless, impenitent, desperate lying! Why, you and I believed him to be guilty until—'

'I beg your pardon! I beg your pardon!' interjected Mr Cupples, laying down his knife and fork. 'I was most careful, when we talked it all over the other night, to say nothing indicating such a belief. *I* was always certain that he was innocent.'

'You said something of the sort at Marlowe's just now. I wondered what on earth you could mean. Certain that he was innocent! How can you be certain? You are generally more careful about terms than that, Cupples.'

'I said "certain",' Mr Cupples repeated firmly.

Trent shrugged his shoulders. 'If you really were, after reading my manuscript and discussing the whole thing as we did,' he rejoined, 'then I can only say that you must have totally renounced all trust in the operations of the human reason; an attitude which, while it is bad Christianity and also infernal nonsense, is oddly enough bad Positivism too, unless I misunderstand that system. Why, man—'

'Let me say a word,' Mr Cupples interposed again, folding his hands above his plate. 'I assure you I am far from abandoning reason. I am certain he is innocent, and I always was certain of it, because of something that I know, and knew from the very beginning. You asked me just now to imagine myself on the jury at Marlowe's trial. That would be an unprofitable exercise of the mental powers, because I know that I should be present in another capacity. I should be in the witness-box, giving evidence for the defence. You said just now, "If there were a single piece of evidence in support of his tale." There is, and it is my evidence. And,' he added quietly, 'it is conclusive.' He took up his knife and fork and went contentedly on with his dinner.

The pallor of sudden excitement had turned Trent to marble while Mr Cupples led laboriously up to this statement. At the last word the blood rushed to his face again, and he struck the table with an unnatural laugh. 'It can't be!' he exploded. 'It's something you fancied, something you dreamed after one of those debauches of soda and milk. You can't really mean that all the time I was working on the case down there you knew Marlowe was innocent.'

Mr Cupples, busy with his last mouthful, nodded brightly. He made an end of eating, wiped his sparse moustache, and then leaned forward over the table. 'It's very simple,' he said. 'I shot Manderson myself.'

'I am afraid I startled you,' Trent heard the voice of Mr Cupples say. He forced himself out of his stupefaction like a diver striking upward for the surface, and with a rigid movement raised his glass.

But half of the wine splashed upon the cloth, and he put it carefully down again untasted. He drew a deep breath, which was exhaled in a laugh wholly without merriment. 'Go on,' he said.

'It was not murder,' began Mr Cupples, slowly measuring off inches with a fork on the edge of the table. 'I will tell you the whole story. On that Sunday night I was taking my before-bedtime constitutional, having set out from the hotel about a quarter past ten. I went along the field path that runs behind White Gables, cutting off the great curve of the road, and came out on the road nearly opposite that gate that is just by the eighth hole on the golf-course. Then I turned in there, meaning to walk along the turf to the edge of the cliff, and go back that way. I had only gone a few steps when I heard the car coming, and then I heard it stop near the gate. I saw Manderson at once. Do you remember my telling you I had seen him once alive after our quarrel in front of the hotel? Well, this was the time. You asked me if I had, and I did not care to tell a falsehood.'

A slight groan came from Trent. He drank a little wine, and said stonily, 'Go on, please.'

'It was, as you know,' pursued Mr Cupples, 'a moonlight night, but I was in shadow under the trees by the stone wall, and anyhow they could not suppose there was any one near them. I heard all that passed just as Marlowe has narrated it to us, and I saw the car go off towards Bishopsbridge. I did not see Manderson's face as it went, because his back was to me, but he shook the back of his left hand at the car with extraordinary violence, greatly to my amazement. Then I waited for him to go back to White Gables, as I did not want to meet him again. But he did not go. He opened the gate through which I had just passed, and he stood there on the turf of the green, quite still. His head was bent, his arms hung at his sides, and he looked some-how—rigid. For a few moments he remained in this tense attitude, then all of a sudden his right arm moved swiftly, and his hand was at the pocket of his overcoat. I saw his face raised in the moonlight, the teeth bared, and the eyes glittering, and all at once I knew that the man was not sane. Almost as quickly as that flashed across my mind, something else flashed in the moonlight. He held the pistol before him, pointing at his breast.

'Now I may say here I shall always be doubtful whether Manderson really meant to kill himself then. Marlowe naturally thinks so, knowing nothing of my intervention. But I think it quite likely he only meant to wound himself, and to charge Marlowe with attempted murder and robbery.

'At that moment, however, I assumed it was suicide. Before I knew what I was doing I had leapt out of the shadows and seized his arm. He shook me off with a furious snarling noise, giving me a terrific blow in the chest, and presenting the revolver at my head. But I seized his wrists before he could fire, and clung with all my strength—you remember how bruised and scratched they were. I knew I was fighting for my own life now, for murder was in his eyes. We struggled like two beasts, without an articulate word, I holding his pistol-hand down and keeping a grip on the other. I never dreamed that I had the strength for such an encounter. Then, with a perfectly instinctive movement—I never knew I meant to do it—I flung away his free hand and clutched like lightning at the weapon, tearing it from his fingers. By a miracle it did not go off. I darted back a few steps, he sprang at my throat like a wild cat, and I fired blindly in his face. He would have been about a yard away, I suppose. His knees gave way instantly, and he fell in a heap on the turf.

'I flung the pistol down and bent over him. The heart's action ceased under my hand. I knelt there staring, struck motionless; and I don't know how long it was before I heard the noise of the car returning.

'Trent, all the time that Marlowe paced that green, with the moonlight on his white and working face, I was within a few yards of him, crouching in the shadow of the furze by the ninth tee. I dared not show myself. I was thinking. My public quarrel with Manderson the same morning was, I suspected, the talk of the hotel. I assure you that every horrible possibility of the situation for me had rushed across my mind the moment I saw Manderson fall. I became cunning. I knew what I must do. I must get back to the hotel as fast as I could, get in somehow unperceived, and play a part to save myself. I must never tell a word to any one. Of course I was assuming that Marlowe would tell every one how he had found the body. I knew he would suppose it was suicide; I thought every one would suppose so.

'When Marlowe began at last to lift the body, I stole away down the wall and got out into the road by the clubhouse, where he could not see me. I felt perfectly cool and collected. I crossed the road, climbed the fence, and ran across the meadow to pick up the field path I had come by that runs to the hotel behind White Gables. I got back to the hotel very much out of breath.'

'Out of breath,' repeated Trent mechanically, still staring at his companion as if hypnotized.

'I had had a sharp run,' Mr Cupples reminded him. 'Well, approaching the hotel from the back I could see into the writing-room through the open window. There was nobody in there, so I climbed over the sill, walked to the bell and rang it, and then sat down to write a letter I had meant to write the next day. I saw by the clock that it was a little past eleven. When the waiter answered the bell I asked for a glass of milk and a postage stamp. Soon afterwards I went up to bed. But I could not sleep.'

Mr Cupples, having nothing more to say, ceased speaking. He looked in mild surprise at Trent, who now sat silent, supporting his bent head in his hands.

'He could not sleep,' murmured Trent at last in a hollow tone. 'A frequent result of over-exertion during the day. Nothing to be alarmed about.' He was silent again, then looked up with a pale face. 'Cupples, I am cured. I will never touch a crime-mystery again. The Manderson affair shall be Philip Trent's last case. His high-blown pride at length breaks under him.' Trent's smile suddenly returned. 'I could have borne everything but that last revelation of the impotence of human reason. Cupples, I have absolutely nothing left to say, except this: you have beaten me. I drink your health in a spirit of self-abasement. And *you* shall pay for the dinner.'

OXFORD

MORE OXFORD PAPERBACKS

This book is just one of nearly 1000 Oxford Paperbacks currently in print. If you would like details of other Oxford Paperbacks, including titles in the World's Classics, Oxford Reference, Oxford Books, OPUS, Past Masters, Oxford Authors, and Oxford Shakespeare series, please write to:

UK and Europe: Oxford Paperbacks Publicity Manager, Arts and Reference Publicity Department, Oxford University Press, Walton Street, Oxford OX2 6DP.

Customers in UK and Europe will find Oxford Paperbacks available in all good bookshops. But in case of difficulty please send orders to the Cash-with-Order Department, Oxford University Press Distribution Services, Saxon Way West, Corby, Northants NN18 9ES. Tel: 0536 741519; Fax: 0536 746337. Please send a cheque for the total cost of the books, plus £1.75 postage and packing for orders under £20; £2.75 for orders over £20. Customers outside the UK should add 10% of the cost of the books for postage and packing.

USA: Oxford Paperbacks Marketing Manager, Oxford University Press, Inc., 200 Madison Avenue, New York, N.Y. 10016.

Canada: Trade Department, Oxford University Press, 70 Wynford Drive, Don Mills, Ontario M3C 1J9.

Australia: Trade Marketing Manager, Oxford University Press, G.P.O. Box 2784Y, Melbourne 3001, Victoria.

South Africa: Oxford University Press, P.O. Box 1141, Cape Town 8000.

OXFORD POETS

A PORTER SELECTED

Peter Porter

This selection of about one hundred of Porter's best poems is chosen from all his works to date, including his latest book, *Possible Worlds*, and *The Automatic Oracle*, which won the 1988 Whitbread Prize for Poetry.

What the critics have said about Peter Porter:

'I can't think of any contemporary poet who is so consistently entertaining over such a variety of material.' John Lucas, *New Statesman*

'an immensely fertile, lively, informed, honest and penetrating mind.' Stephen Spender, *Observer*

'He writes vigorously, with savage erudition and wonderful expansiveness . . . No one now writing matches Porter's profoundly moral and cultured overview.' Douglas Dunn, *Punch*

OXFORD POETS

FLEUR ADCOCK

Time Zones

In this lively new collection, Fleur Adcock's subjects range from domestic matters—recalling the birth of her son some years back; remembering her father, the news of whose death in New Zealand reaches her, the expatriate, in England; working in her own London garden—to matters of contemporary concern, such as the Romanian bid for freedom in 1989, and support for Green causes, including the anti-nuclear stand.

'She is an eminently readable poet, whose quiet accuracy sometimes makes me laugh out loud.'
Wendy Cope, *Guardian*

THE WORLD'S CLASSICS
THE WIND IN THE WILLOWS
Kenneth Grahame

The Wind in the Willows (1908) is a book for those 'who keep the spirit of youth alive in them; of life, sunshine, running water, woodlands, dusty roads, winter firesides'. So wrote Kenneth Grahame of his timeless tale of Toad, Mole, Badger, and Rat in their beautiful and benevolently ordered world. But it is also a world under siege, threatened by dark and unnamed forces—'the Terror of the Wild Wood' with its 'wicked little faces' and 'glances of malice and hatred'—and defended by the mysterious Piper at the Gates of Dawn. *The Wind in the Willows* has achieved an enduring place in our literature: it succeeds at once in arousing our anxieties and in calming them by giving perfect shape to our desire for peace and escape.

The World's Classics edition has been prepared by Peter Green, author of the standard biography of Kenneth Grahame.

'It is a Household Book; a book which everybody in the household loves, and quotes continually; a book which is read aloud to every new guest and is regarded as the touchstone of his worth.' A. A. Milne

OXFORD BOOKS

THE OXFORD BOOK OF ENGLISH GHOST STORIES

Chosen by Michael Cox and R. A. Gilbert

This anthology includes some of the best and most frightening ghost stories ever written, including M. R. James's 'Oh Whistle, and I'll Come to You, My Lad', 'The Monkey's Paw' by W. W. Jacobs, and H. G. Wells's 'The Red Room'. The important contribution of women writers to the genre is represented by stories such as Amelia Edwards's 'The Phantom Coach', Edith Wharton's 'Mr Jones', and Elizabeth Bowen's 'Hand in Glove'.

As the editors stress in their informative introduction, a good ghost story, though it may raise many profound questions about life and death, entertains as much as it unsettles us, and the best writers are careful to satisfy what Virginia Woolf called 'the strange human craving for the pleasure of feeling afraid'. This anthology, the first to present the full range of classic English ghost fiction, similarly combines a serious literary purpose with the plain intention of arousing pleasing fear at the doings of the dead.

'an excellent cross-section of familiar and unfamiliar stories and guaranteed to delight' *New Statesman*

ILLUSTRATED HISTORIES IN OXFORD PAPERBACKS

THE OXFORD ILLUSTRATED HISTORY OF ENGLISH LITERATURE

Edited by Pat Rogers

Britain possesses a literary heritage which is almost unrivalled in the Western world. In this volume, the richness, diversity, and continuity of that tradition are explored by a group of Britain's foremost literary scholars.

Chapter by chapter the authors trace the history of English literature, from its first stirrings in Anglo-Saxon poetry to the present day. At its heart towers the figure of Shakespeare, who is accorded a special chapter to himself. Other major figures such as Chaucer, Milton, Donne, Wordsworth, Dickens, Eliot, and Auden are treated in depth, and the story is brought up to date with discussion of living authors such as Seamus Heaney and Edward Bond.

'[a] lovely volume . . . put in your thumb and pull out plums' Michael Foot

'scholarly and enthusiastic people have written inspiring essays that induce an eagerness in their readers to return to the writers they admire' *Economist*

OXFORD REFERENCE

THE CONCISE OXFORD COMPANION TO ENGLISH LITERATURE

*Edited by Margaret Drabble and
Jenny Stringer*

Based on the immensely popular fifth edition of the
Oxford Companion to English Literature this is an
indispensable, compact guide to the central matter
of English literature.

There are more than 5,000 entries on the lives
and works of authors, poets, playwrights, essayists,
philosophers, and historians; plot summaries of
novels and plays; literary movements; fictional
characters; legends; theatres; periodicals; and much
more.

The book's sharpened focus on the English literature of the British Isles makes it especially convenient to use, but there is still generous coverage of
the literature of other countries and of other disci-
plines which have influenced or been influenced by
English literature.

From reviews of *The Oxford Companion to
English Literature*:

'a book which one turns to with constant pleasure
. . . a book with much style and little prejudice' Iain
Gilchrist, *TLS*

'it is quite difficult to imagine, in this genre, a more
useful publication' Frank Kermode, *London Review of Books*

'incarnates a living sense of tradition . . . sensitive
not to fashion merely but to the spirit of the age'
Christopher Ricks, *Sunday Times*

OXFORD POPULAR FICTION

THE ORIGINAL MILLION SELLERS!

This series boasts some of the most talked-about works of British and US fiction of the last 150 years—books that helped define the literary styles and genres of crime, historical fiction, romance, adventure, and social comedy, which modern readers enjoy.

Riders of the Purple Sage	Zane Grey
The Four Just Men	Edgar Wallace
Trilby	George Du Maurier
Trent's Last Case	E C Bentley
The Riddle of the Sands	Erskine Childers
Under Two Flags	Ouida
The Lost World	Arthur Conan Doyle
The Woman Who Did	Grant Allen

Forthcoming in October:

Olive	Dinah Craik
The Diary of a Nobody	George and Weedon Grossmith
The Lodger	Belloc Lowndes
The Wrong Box	Robert Louis Stevenson